BAMBOOS

Christine Recht and Max F. Wetterwald

edited by *David Crampton*

translated from the German by *Martin Walters*

B. T. Batsford Ltd · London

Frontispiece:
'Grass, bamboos, orchids and rocks',
Chinese fabric-drawing, Lu Kunfeng (1980)

This edition first published in 1992
Reprinted 1993/1994/1995/1996/1998/1999/2000
First published in paperback 2001, reprinted 2002
Translation © B.T. Batsford Ltd 1992
German edition © 1988 by
Eugen Ulmer GmbH & Co.,
Stuttgart, Germany

B.T.Batsford Ltd
64 Brewery Road
London N7 9NY

A member of the Chrysalis Group plc.

Typeset by Lasertext Ltd., Stretford
Manchester
and printed in Singapore

ISBN 0-7134-8714-3

Contents

Foreword

Bamboos are fascinating plants – fascinating in their beauty and elegance, in their varied shapes and in their unusual qualities. Bamboos are simultaneously hard and soft, their stems are straight and yet flexible, their leaves graceful and green all year. Small wonder that these plants, which are already in vogue in America, are getting more and more popular in Europe.

Most bamboos come from Asia. There they are important in everyday life. All kinds of use is made of them – their shoots are eaten and bamboo groves are a feature of the natural landscape. Bamboos also influence the art and culture of many Asian people. In China they are seen as the embodiment of the Chinese way of life: yielding, yet always victorious.

In Europe, bamboo is mainly a decorative plant, an exotic that survives our climate and fits into the landscape. Many species are frost-tolerant and decorate our gardens in winter with their soft foliage. Bamboos can be used in so many different ways, like scarcely any other group of plants. They fit into both small and large gardens, where they can grow in groves, as a hedge, and also as individual plants. They can therefore be used both in a supporting role and as a main performer in the garden. They are effective on terraces, balconies, in conservatories, and even on roofs.

More and more nurseries offer container-grown bamboos. Some outlets have specialized in bamboo propagation and now have a hundred or more different species available. Bamboos are very beautiful but until recently we have not known how to deal with them effectively. Fascination with bamboos has led many gardeners to buy them, but ignorance of their requirements has often brought frustration and despair.

This book will therefore help all bamboo enthusiasts. It will explain how to treat these magical plants and point out which sites suit particular bamboo species well and where they look best. It will increase understanding of these fine plants and enable gardeners to succeed with them by understanding their particular requirements.

When writing about bamboos one feels rather like the bamboo painters of ancient China who said: 'If you want to paint a bamboo you have to become one of their kind.' It is particularly difficult to write a book about bamboos in Germany and to collect together the basic information, because there are so few experts.

I should like to thank all those who have helped me with their knowledge, experience and passion for bamboos: Werner Simon, Marktheidenfeld, who as co-author collated and described all the species that can be cultivated in Germany; Albrecht Weiß from Seeheim-Jungenheim, who put his extensive knowledge and long experience of bamboos at my disposal and who also infected me with his enthusiasm, as did Ullrich Willumeit, Darmstadt. Thanks also to Dr. Warda of the New Botanic Garden, Hamburg where Max Wetterwald was able to photograph bamboos, and to Wolfgang Eberts, Baden-Baden, who always knew the answers to particular problems as well as the most beautiful bamboo gardens to photograph.

Christine Recht

1

Bamboos in Asiatic Culture

'Bamboo is my brother', so goes a Vietnamese proverb. This demonstrates precisely the relationship of almost all Asiatic people, even today, with this plant. From China to India, from the moist, primeval forests to the cool mountain foothills, bamboo is such a natural partner to humans in all walks of life that to live without it is scarcely imaginable. 'It is quite possible not to eat meat, but not to be without bamboo', so said Su Dongpo (1036–1101), the greatest poet and artist of the Song Dynasty. It goes deeper than this, however. In ancient China people identified with bamboo, the symbol of the Chinese way of life. Bamboo stands for elasticity, endurance and tenacity. Bamboo bends in the breeze, but does not break. The leaves are moved by the wind, but do not fall. Bamboo therefore survives and conquers.

In Japan this characteristic is still known as a 'bamboo mentality': to make compromise, to yield, and eventually to go forward unbroken from all contests. In Asia bamboo embodies the ideas of Taoism, laid down mainly by Laotse. These ideas describe the art of survival as yielding and then coming back again.

Religion and symbolism

In all Asiatic countries religion is closely connected with nature. The gods, of which there were and still are countless numbers, lived in the rocks, in the water, woods, and hedgerows. Individual stones, rivers and trees were also held to be holy, although not in the European sense of the word. Water, mountains, plants and animals were seen not as inferior to humans but as part of a whole, to which people also belong. The belief that humanity cannot exist without nature, and that we must live by its laws in order to survive is still held in modern Asia. With this appreciation of the unity of humanity and nature one can understand why in Asia bamboo is considered as a friend, or a travelling companion.

It is not only bamboo that plays a leading role in Asiatic religion, philosophy and art; pine, willow, plum, lotus flower and chrysanthemum are also important. But of all these symbolic and revered plants, it is the bamboo alone that is put to practical use, be it as building material, food, or for making thousands of objects in daily use. Bamboo is of special significance in Asiatic, and particularly Chinese, symbolic language. It droops its leaves because its inner self (that is its heart) is empty. An empty heart in China means modesty, so bamboo is a symbol for this virtue. Bamboo is evergreen and does not change its appearance with the seasons. It is therefore held as a symbol of age. The Chinese character for bamboo is similar to that for laughter, because the Chinese believe that the bamboo plants bends when it hears laughter. In Chinese, the words for bamboo, prayer and wish all sound the same. The reason is as follows: sticks of bamboo explode if placed in a fire, with a loud crack. The bamboo firework was supposed to drive away demons and ensure that the gods heard prayers and wishes for peace and tranquillity.

In Asian art bamboo is often illustrated together with orchids or plum blossom. The flowers embody woman – or yin, the female element, bamboo the man – or yang, the male element.

Because bamboo plays such an important role in people's lives in Asia, in philosophy as well as in everyday practical life, it has become a part of legend, belief and superstition. There are countless fairy-tales and legends in Asia concerning bamboos. Here is an example: When a Japanese farmer was cutting bamboo he found a tiny girl

inside a bamboo culm. He took her home and brought her up as his own child. She grew up to be one of the most beautiful and charming girls in the whole country. The emperor of Japan heard about her and wanted to marry her. However, the girl wrote him a letter saying that it was too great an honour for her and she decided to return to the bamboo. The emperor sent all his soldiers to find her, but without success, and in sadness at losing the bamboo girl he burned her letter on top of the mountain, Fujiyama, where the fire still burns to this day.

Many Asian ceremonies, especially in Japan, are closely connected with bamboo. For example, only certain species of bamboo are used for making the equipment so important in the Japanese tea ceremony. In certain parts of Japan there is also the 'bamboo splitting festival' which has been going since the eighth century. It is opened by priests with purifying ceremonies before the young men of the village start to split the fresh bamboo canes.

Traditional new year decorations in a Japanese house involve the three most revered plants: bamboo, plum and pine. These plants are known as the 'three friends' and they symbolize the three religious writers: Buddha (bamboo), Confucius (plum) and Lao Tse (pine). In China 'four noble plants' are respected: bamboo, orchid, plum and chrysanthemum. Together they stand for good luck and well-being and are present at all festivals, be these of a secular or religious nature. At the birth of a child, if the umbilical cord is cut with a bamboo knife this indicates a life of happiness. In Japan this privilege was earlier reserved only for the children of the 'god-like'.

In Asia gods have human customs and requirements and are often depicted with very familiar everyday objects. Thus the immortal Ho sien-Ku is cooking rice with a bamboo spoon in his hand even as he is 'redeemed and rises into the air'. He can be seen in old paintings with this bamboo spoon is his hand.

(*Left*) Thick bamboo forest of *Dendrocalamus giganteus* in Cibodas (Java)

Bamboo brush and bamboo paper

'His name may be written on bamboo and silk' – so runs one of the many Chinese sayings. Bamboo has played a major role in calligraphy and artistic writing from very early times and calligraphy is still a high art form today. The pens used now, as in ancient times, to paint the characters are cut from bamboo stems. Paper used to be made from bamboo leaves and the shape of the leaf is still influential in Asian calligraphic work.

Calligraphy should not be seen simply as fine writing but as an art form that reached its highest development centuries ago, particularly in China. 'Writing means picture painting and painting means picture writing', so it is said in Asia. A calligrapher was therefore not simply a well-educated person who wrote text, but gave the text a particular quality through the artistic aspects of the script. It is possible to appreciate Chinese characters for their aesthetic attraction alone but the pleasure is deepened if one can also read the words, since content and shape are closely connected. The first recognizable Chinese characters are found in the thirteenth century BC. By then they were already simplified drawings reduced, so to speak, to shorthand. And so it is still today. Over the millenia there have been a series of calligraphic 'schools', each with its own famous master, who was both poet and artist.

Splitting bamboo culms for painting and wickerwork

Many stories surround each of these, for example the story of the Buddhist monk Huaisu who lived around the year AD 725. He was so obsessed with his art that he painted characters on every surface he came across: temple walls, pieces of clothing, pots and pans. He even grew banana trees so he could draw on their large leaves. His eccentric style was described as comparable to 'frightened snakes and sudden storms'. By contrast, the calligraphy of Wang Xizhi (307–65) was compared to mist and settling dew.

Eastern calligraphy takes its particular artistic form from the brush, known in China for about 6,000 years. From the beginning this was finely constructed: animal hair on a long, flexible bamboo handle. The hair was arranged in a wedge shape – thick at the base and increasingly thin towards the tip – so that both the finest lines and broad, powerful strokes could be painted. This brush shape has changed very little to this day. Just as unusual is the way in which the brush is used. It is held between the fingers, but the fingers and hand are not moved when writing or painting. The artist draws and paints from the elbow and shoulder, so that what appears on the paper comes direct from the centre of the body; that is from the heart. 'The brush dances and the ink sings' is still said of fine calligraphy and ink drawings. Whilst the brush is still made of bamboo today, in much earlier times so was the writing surface – thin bamboo blocks. These blocks were bound together to make the first 'books'. Later on, painting and writing was done mostly on silk, until this ran out because of the high demand (China had special offices where writing and copying was carried out). Paper was discovered in China about 2,000 years ago. In 105 BC the court eunuch Tsai Lun told the emperor about a new discovery. A paste was prepared from old fishing nets, tree bark, hemp and grasses and spread out thinly on mats to dry. Archaeological remains show that paper had been made in a similar way from bamboo leaves at least two

Bamboo paintings, on a scroll (*left*) and on rice paper (*right*), a centuries-old tradition

centuries earlier. The bamboo leaves were soaked in water and beaten all day until a thin paste resulted, which was poured out on to mats and dried. This gave thin sheets of paper which were then carefully smoothed out. Bamboo paper, however, was somewhat fibrous and the ink tended to run; therefore more and more paper was produced from waste. This, as well as being cheaper, is perfectly suited to writing and drawing with ink.

With the discovery of paper the art of calligraphy really took off, first in China and later also in Japan. As well as the artists there were many copiers who copied famous and beautiful works of art and this work was held to be desirable and honourable. In AD 400 calligraphy was taken from China to Japan, and many scholars and artists fled to Japan and settled there. Japanese artists refined calligraphy almost to the point of abstraction. There was also a simpler script that was felt by the artists to be too 'primitive' for them. This was taken up by many noble ladies, which explains why many famous works of Japanese literature were written by women.

First in China, then in Japan and other Asiatic countries, painting developed in parallel with calligraphy. A well-written text is often accompanied by symbolic ink drawings which relate to the text. The large, often metre-long, hanging pictures often have an explanatory text or poem in fine calligraphy. In Asia, calligraphy, painting and poetry are not considered separate arts as they are in the West, but aspects of one art. A poet is a painter and calligrapher; a painter is also a person of letters.

The bamboo painter

Bamboo has long been portrayed on ink drawings and scroll paintings. This is because bamboo is characteristic of the Asiatic landscape, and nature is still the most popular subject chosen by artists, along with religious and courtly scenes. In China, nature is not presented in simple form; the artists wanting their pictures to express much more. A realistically painted landscape, however fine, is regarded as primitive, and the artists put much more into the composition. In Japan, by contrast, landscapes tend to be more realistic and colourful, but bamboo is still dominant because it is such a prominent part of the natural scene.

On the other hand, bamboo is also used purely symbollically, for example as indicative of the Taoist concept of yielding in order to overcome. The portrayal of bamboo, orchid, plum blossom and chrysanthemum (the four noble plants) and lotus and pine developed into a particular genre, held above all others, in the Yuang period of China. There were special bamboo painters who concentrated on the representation of bamboo. Three of the most famous were Zhao Mengfu, his wife Kuang, and Gao Kegong – all living around AD 1250. These artists, like many others, accompanied their ink drawings with poetic verses. The drawings, mostly in black ink, or more rarely painted with blue-green, are very spiritual, capturing the character of bamboo through concise suggestion and they have great radiance. 'In order to paint bamboo you must become bamboo' say the bamboo artists.

There were also specialist bamboo painters in Japan. They simplified the themes still further, following Zen philosophy, emphasizing the symbolism and cleverly refining their paintings following the lifestyle of the Japanese imperial court.

Bamboo poets

Just as scroll paintings and ink drawings became increasingly simplified and symbolic, so too did poetry. In Japan, Haiku is still popular. This is a rhythmic three-line poem that expresses a particular atmosphere in few words. A Haiku poem describes, as it were, the scene presented by an instant, yet incorporating past and future. Naturally the ever-present bamboo features in many Haiku pieces.

Gardens in China and Japan

Bamboo plays an important role in Chinese and Japanese gardens. To understand this, one needs to appreciate the significance of gardens and garden art in these countries. Garden design is considered an art form similar to landscape painting, and is carried out according to a comparable set of rules. (There are fundamental differences between Chinese and Japanese garden art, but it would take us too far to go into these here. The reader should investigate the extensive literature.) In Asia there has always been a particular fear of wild, untamed nature, yet simultaneously a strong love for it and a feeling of unity with it. Nature and humanity are seen as indivisible. Gardens should reflect nature, but also provide people with the chance to immerse themselves in nature through meditation. A garden is therefore strongly symbolic. Chinese gardens are works of art in which landscapes are set up, not to imitate nature, but to simplify it and render it more profound, as in painting. A Chinese or Japanese garden is scarcely comparable with a European garden, because it is based upon a different set of assumptions. Chinese and Japanese gardens create a landscape that stimulates the imagination rather than the understanding. One sees over and over the juxtaposition of Yang and Yin, the masculine and feminine, hard and soft; for example rock and water, bam-

'Two waders after rain', ink drawing
by Lu Kunfeng (1980)

boo and chrysanthemum, straight and curved lines.

In 1634 Yuan Jeh wrote this about gardens: 'A single mountain can have many effects, a small stone can awaken many feelings. The shadows of dry banana leaves draw themselves wonderfully upon the paper of the window. The roots of the pine tree force themselves through the cracks of the stone. If you can find peace here in the middle of the city why should you wish to leave this place and seek another? . . .' All things in Chinese gardens – and in still more refined and abstract form in Japanese gardens – have symbolic value and are aids to meditation. Water is always present, standing for human life and philosophical thought. There are no lawns, but gravel beds instead. Rocks symbolize mountains. They are often raised up high and particular value is laid on bizarre and steep formations. They represent, in contrast to water, the might of nature. Flowers are never planted in groups or patterns but stand isolated, to aid meditation. The chrysanthemum, which flowers late and is frost-tolerant, symbolizes culture and retirement, the water lily is the sign of purity and truth. Bamboo stands for suppleness and power, true friendship and vigorous age. The evergreen foliage of bamboo also provides a background for plum blossom and makes an artistic picture together with pine. In Asian gardens bamboo is usually thinned so that individual

(*Left*) Bamboo, water and pagoda roofs, a place for meditation. Daguanlou Park, Kunming, China

stems can be seen clearly, and a bamboo with many stems represents an entire forest.

In Japan, gardens are arranged so that they relate to living quarters. The sliding panels of traditional Japanese houses open onto the garden and the garden is designed so that when the panel is opened it appears like a painting, in which deep meditation is possible.

In the sixteenth century Japanese gardeners refined their art so much that gardens were laid out with scarcely any plants at all. A rock symbolized a mountain or waterfall. The water was indicated by fine sand or gravel, raked to give patterns symbolizing a flowing stream, surging river or open sea. Such dry gardens can still be seen today and they often contain just a few particularly fine bamboo stems. They are not walked in but are simply admired from the house.

Artistic gardens were laid out in temple grounds. The bamboo hedge of Hokokuji Temple in Kamakura is famous. Here, individual moss-covered stones and stone lanterns lie between giant bamboos. Bamboo hedges round temples contain particularly rare or fine bamboo species. In Kochi in Japan there is a hedge of the rare 'golden bamboo' which has golden-yellow stems with green stripes. This has been placed under special protection as a natural monument. Hedges of 'tortoise-shell' bamboo are also well known and often visited.

2

The Uses of Bamboo

Bamboo is a plant which accompanies people throughout their lives in Asia and for this reason it plays a big role in Asiatic culture. Asian people, more than half the population of the world, use bamboo for building their houses, as containers for food and drink, for making all sorts of household and agricultural implements, for making weapons, as food, animal fodder, and as medicine.

The British Colonel Barrington de Fonblanque, who travelled in China in the last century, summarized his impressions thus: 'How could the poor Chinese survive without bamboo? It supplies not only nourishment, but also the roof of his house, the mat on which he sleeps, the cup from which he drinks, and the chopsticks with which he eats. The Chinaman waters his fields using bamboo pipes, gathers in his harvest with a bamboo rake, purifies the grain with a bamboo riddle and carries it home in a bamboo basket. The mast of his junk is made from bamboo, as is the axle of his barrow. He is whipped with a bamboo rod, tortured with bamboo splinters and eventually strangled with a bamboo rope.'

The importance of bamboo in Asia has not been diminished by the impact of modern technology. The qualities of this plant are such that it cannot very often be replaced, even with modern substitutes. A bamboo pipe is as strong as steel, but more malleable. It grows tall and straight and is therefore perfect as a building material. It is light and hard, flexible and tough. It is difficult to set on fire and can be bent under heat, retaining its shape without losing its strength and elasticity. It can be split into the finest fibres, from which an almost unbreakable kind of rope can be made. The shoots are a vitamin-rich delicacy and the leaves provide nourishing animal fodder.

Thick bamboo forests are an integral feature of every Asian landscape, from Japan to India, from the equator northwards to 40 degrees latitude, from tropical rain forest to the cool foothills of the Himalayas. The dry rustle of their narrow leaves is, together with the fizzing of cicadas and the noise of frogs croaking, the characteristic sound of a tropical evening.

Cultivation in commerce and agriculture

There are 1,050 to 1,070 different species of bamboo, from the tiny species of *Sasa* to gaint subtropical bamboos up to 30 metres tall with a diameter of 30 cm (12 in). Only a few species are cultivated in commercial plantations. In Japan, where there are about 100 species, only about 15 species are widespread and in cultivation. Ninety percent of all bamboo plantations in the land of the rising sun consist of 'Madake' (*Phyllostachys bambusoides*) and 'Moso' (*P. heterocycla f. pubescens*). Madake is mainly used for building, Moso for food, since its shoots are considered the finest of all.

It is a similar story in all other Asiatic countries: only the few species best suited to the local conditions are used. These can sometimes be 'imported' species. Moso and Madake, for example, originate in China but were long ago introduced to Japan simply because they were – and still are – the best species for these purposes. Asiatic people are realists; they do not experiment much with bamboos, but plant them in places where the conditions are traditionally known to be optimal. The plantations are carefully tended and the bamboo culms harvested at the correct time – when they are three to five years old. Before that the culms are too soft, later on they get too hard. Even in areas where bamboo is not cultivated, but where it grows naturally over large

Tropical bamboo in cultivation

areas, Asian people treat it with care. They know that clearing it is not sensible because it takes years to grow back to productive size. They therefore harvest only those culms that are ready for cutting.

In spite of this, large bamboo groves are in danger of destruction because of the pressures of rapid population growth. Tall bamboo species grow on very fertile soil and many villages have extended their agricultural land into sites of large bamboo stands. This is also one of the reasons why the Giant Panda is threatened. This large animal, with the deceptively harmless look of a children's toy, feeds almost exclusively on bamboo. For centuries it has been able to find 20 different species in its environment. Some grew in the warm lowlands, others in the cool, misty mountains – prime panda country. Pandas could change their habitat and still have no trouble finding adequate food, However, now that bamboo forests have been increasingly converted to agriculture, both in the lowland and mountain foothills, there is a smaller range of species available. When some of these flower together and die

Cut culms of *Dendrocalamus giganteus* in Java

Bamboo scaffolding in Delhi

off, or are at least considerably reduced for several years, the pandas cannot get enough food and therefore starve. People have tried, without success, to accustom pandas to eating rice and grasses. Many pandas have died with the flowering of the bamboos and those that survive are also in danger because it takes decades for suitable bamboo stands to re-establish, either from the few remaining rhizomes, or from seed.

Building with bamboo

For people, however, the reservoir of bamboo in Asia is almost inexhaustible and it will remain so since the plant is too closely bound up with human life to be dispensed with. One sees the importance of the plant particularly clearly in isolated areas, for example in the villages of northern Thailand or Laos. There, far from civilization, there are self-sufficient communities. In these Meo or Lao villages the versatility of bamboo is particularly noticeable. It is not only most of the household utensils that are made of bamboo, but also water conduits from the local stream, the flute for evening music-making, the water-pipe, bow and arrow, the fence of the pig enclosure, and the palisade surrounding the village. Bamboo provides protection from monsoon storms and from the cold at night. The base of the house is thick bamboo stems rammed into the ground. On that rests the floor of bamboo slats, a metre (yard) or

Bamboo ladder against a house in Sumatra

and concrete. However, whenever the walls need cleaning bamboo scaffolding is used, held together by bamboo rope and reaching to the highest floors. Occasionally one sees, perhaps at the 45th storey, a small, hanging bamboo scaffold, for a bit of local repair work.

Of course, familiar steel scaffolding is also used in Asian cities. But builders prefer bamboo, and not simply because of tradition. Bamboo, unlike iron and steel does not rust in the damp tropical climate. Bamboo stems are lighter, but still very hard. Whenever a typhoon has hit Hong Kong ones sees the same picture; the few steel scaffoldings lie like matchsticks on the ground while the bamboo structures are still in place, if a little bent in places. They can be, however, repaired in less than a day.

Bamboo has other uses in small Asian building sites, too. Bamboo stems are used to make shutters. Since human labour is cheaper than using machines, soil is often moved by hand; it is dug out by broad spades and loaded into flat bamboo baskets that are carried away by women.

Bamboo implements

Bamboo is a suitable material for all kinds of containers in Asia – from delicately curved beakers and cups to large nets, woven from rough rope, in which pigs are transported to market. Women fetch water from the well in tall bamboo pipes, and charcoal, so vital for making fire for cooking, is stored in short bamboo containers, sealed at the end with loam.

Most widespread, however, is the bamboo basket, found in every shape and size. Whether it is just a simple container for fetching potatoes from market, or a valuable woven jewellery casket, such as Japanese craftsmen have produced following centuries-old tradition, the basic material is always the same – bamboo. The Chinese use portable stacking food containers made of woven bamboo. These are lacquered and painted as works of art. Chinese also use tightly woven bamboo vessels for steaming certain dishes.

more above the ground to protect against snakes and other animals, as well as from flooding. The walls are made from woven bamboo mats and bamboo roller-blinds hang in front of the windows. The palm- or banana-leaf roof rests on a framework of bamboo. Large families live in these light, airy houses, which last for years or even decades. When one does fall into desrepair, it doesn't take more than a few days' work to build a new one.

However, it is not just in the countryside and forest regions that bamboo is used for building. In major cities like Hong Kong or Singapore, which have energetically entered the industrial age, huge skyscrapers are always springing up. The latest machinery is used in their construction, along with modern materials such as glass, steel

Bamboo is also used for tools. All over Asia one sees gleaming polished bamboo poles, used by men and women as a yoke over the shoulders for carrying balanced, often heavy, loads. One often sees a mother on the way to market with a basket of fruit on one side and a child in a flat bamboo basket on the other.

Clearly, bamboo is used for simple tools such as rakes and brooms. It is also used to make fire. Some Asiatic tribes strike sparks from the hard bamboo wood using a piece of china. Others make a piston that exactly fits the diameter of a bamboo pipe. This is then driven hard and long into the pipe until the heated compressed air inside it ignites wood splinters. Elegant Japanese women use artistically painted delicate folding bamboo fans. Ropes of up to 300 m (1,000 ft) length are still made to this day from the same material. Hanging bridges are built using bamboo rope and columns of several hundred men use the same

(*Above*) Small house made entirely of bamboo, Prafrance

(*Right*) Bamboo tower in the 'Phänomena' exhibition, Zürich. Forty-five Chinese used 250 tonnes of bamboo to build this

rope to pull freight junks up the wild rapids of the Yangtse river.

Even in the West, bamboo has unexpected uses. Experts use bamboo styli for playing old gramophone records, and in Edison's first light bulb the filament was a hair-fine carbonized bamboo thread.

Bamboo weapons

It should come as no surprise that people also make weapons from this material. The elastic, tough and straight bamboo culm lends itself, for example, to the production of bows and arrows.

Huntsmen in New Guinea still make their weapons from thin bamboo culms. In Japan, where archery is part of Zen meditation, specialized craftsmen make bows that are larger than a man. Simpler, but more deadly, are the bamboo blowpipes used by the natives of Sumatra to fire poison arrows. Even the advanced American army lost many soldiers to bamboo weapons during the Vietnam War. The Vietkong dug well-camouflaged pits, with sharpened bamboo stakes in the bottom whose points were usually covered with poison.

A less deadly, but extraordinary effective method is used by jungle hunters to keep wild animals away from their camps. Green, freshly cut bamboo stems are laid in circles around the camp fire, at various distances from it so that they are exposed to the heat of the embers but do not catch fire. The heat causes them to explode with a loud noise which frightens away any marauding animal (or person).

No account of bamboo as a weapon is complete without a mention of the subtle form of execution in ancient China. The condemned person was tied down over an emerging bamboo shoot in such a way that they could not move. The shoot grew straight through the person, killing them in gruesome fashion.

Bamboo as food and medicine

Bamboo also supports life. Bamboo shoots are an important food all over Asia. The shoots, which may be as thin as asparagus or as thick as an arm, depending on the species, are harvested just before they emerge from the soil – as with asparagus.

The shoots are boiled, the water drained off and the outer sheaths removed. They are then cut into thin slices or eaten whole. The consistency is that of an apple, the taste that of an artichoke and the nutritional value that of an onion.

All bamboo species have edible shoots but those of Moso are particularly suitable for the kitchen. At markets one can buy bamboo shoots that have been specially grown for cooking. It is only possible to gather wild bamboo shoots in the less accessible and populated areas.

The shoots are also dried or tinned. In this form they can also be purchased in the West from specialist food shops.

Bamboo has been ascribed wonderful medicinal properties. Two egg yolks, cooked slowly in a bamboo stem, absorb the bamboo sap and are effective against asthma, spitting blood and haemorrhoids. The secret medicine 'Tabaschir' has been used for centuries against poisoning, to much ridicule from sceptics. It has now been discovered that it is the silicic acid in bamboo that is effective. Indeed, silicic acid is now produced synthetically and used worldwide to absorb poison in the stomach.

Distribution outside Asia

There are also large natural areas of bamboo in South America and Africa. Of the numerous genera, species and forms that are native to these countries we know only a few in Europe and America: the tropical genus *Bambusa*, the South American *Chusquea* and the Mexican *Otatea*. Many of the other bamboo species of these regions do not grow well outdoors in temperate climates and can only be cultivated with extreme difficulty in the greenhouse.

Bamboo has less economic value in the West, even for the people of those parts of S. America, Africa and Australia where it is native. There are plenty of native trees, and bamboo is not so widely used, although the leaves are sometimes used for the roofs of primitive huts. *Chusquea* is used as fodder in S. America. The tropical genus *Bambusa* is inedible, so bamboos are not used in cooking. In the West, unlike most Asiatic countries, there are no bamboo plantations, and bamboo is merely a constituent of the tropical forest.

These original tropical forests are, however, a veritable 'El Dorado' for the bamboo enthusiast and in them one finds particularly interesting types of bamboo. These bamboo forests are frequently impenetrable because the strong stems stand close together. One also finds bamboos growing on the jungle tree-trunks as climbers.

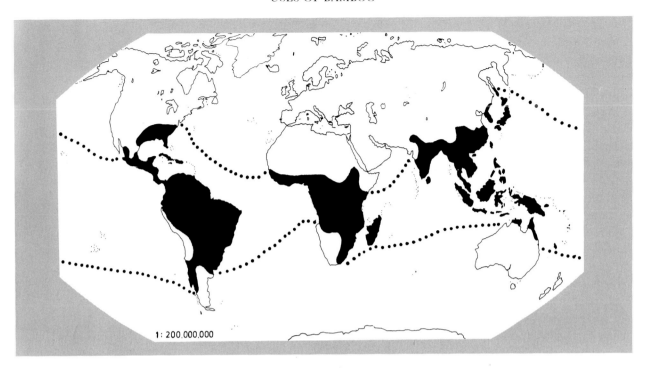

1 : 200,000,000

There are creeping bamboos and even swamp bamboos. There is no doubt that, even though the bamboos of this region are of less value to horticulture, there are species of great interest to botanic gardens and as teaching material to demonstrate the richness and variety to be found in the bamboo family.

Unusual bamboo species from the tropical forests are just beginning to be cultivated, but it is early days yet and it is not possible to say how successful these attempts will prove.

Distribution of bamboo.
Bamboos grow in tropical forests in a broad equatorial belt from 40° S to 40° N, up to 3,000 m (9,800 ft) altitude. They are used and cultivated only in Asia, and it is from this region that most European and American garden species (with a few exceptions) originate. Bamboos have little economic value in Africa and Australia, where there are relatively few species. South America has a wider range of species, including climbing bamboos and epiphytic (tree-living) bamboos

When bamboos came to Europe

Bamboo first came to Europe in the middle of the nineteenth century. Silk importers brought exotic plants back from China and Japan and gave them as presents to noblemen or to their rich clients, or planted them in their own parks. The oldest bamboo stands in Europe are, therefore, found in the gardens of the great estates, or in parks that once belonged to the nobility. The estate of the Markgraf of Baden, in Baden-Baden, West Germany is a good example. Here there is a thriving 80-year-old bamboo grove. European

gardeners knew next to nothing about bamboo and the people who brought the bamboo over on ships knew even less. Therefore in the early days the bamboos planted were clearly those species that survived the transport by boat from the Far East. The tropical species tended not to survive the first winter and more sensitive species did not even survive the journey by ship. This meant that for decades only certain hardy *Phyllostachys* species survived and grew outside their Europe. This was an undirected and entirely accidental selection. In England, with its milder climate and the greater gardening skill of its inhabitants, there were more species in cultivation by the turn of the century than, for example, in Germany.

23

It is possible, however, to trace the entry of bamboo into Europe much further back than to its introduction into cultivation in the last century. The first silk-moth eggs were smuggled out of China to Constantinople in bamboo tubes as early as AD 552. The monks who risked their lives in bringing these pieces of bamboo out of China thus contributed to the eventual loss of significance and decline of the famous Silk Road that for centuries stretched across the whole of Asia. Silk production then began in Europe and China lost its monopoly. European botanists had studied bamboo long before it was introduced into cultivation. In 1626 G. E. Rumpf published a seven-volume work entitled *Herbarium amboinense*, which describes 24 bamboo species. Rumpf called them 'Rohrbäume' (= 'reed-trees'), the name bamboo being used later by Linnaeus (1707–1778), possibly in derivation from the Indian word 'mambu' or 'bambu'. Linnaeus described many bamboo species in his publications. There were also publi-

Phyllostachys heterocycla f. *pubescens* at Cap Ferrat

cations by F. J. Rupprecht (1839) and P. Munro (1868) – descriptions and identification of tropical bamboo genera and species. There is even a short description of Bambusa (bamboo) in Meyer's *Konversations-Lexicon* of 1871.

Interest was then stirred when the first plants came to Europe, began to thrive and found admirers. In 1906 the Belgian botanist Lehaie founded a journal, *Bamboo*, which dealt with attempts to acclimatize bamboo plants in Europe. In 1903 Spörry described 45 genera, and in 1911 France described 230 species. Interest in bamboos was particularly strong in England. This is understandable since English sailors travelled the seas of the world and traded eagerly in the Far East. Gardeners in France also became increasingly interested in bamboos and species could be grown, particularly in the south, that would not survive in the cooler north of Europe.

The interest in bamboos shown by certain enterprising French gardeners led to the development of Prafrance (half an hour from Nîmes) as a place of pilgrimage for bamboo enthusiasts from all over Europe.

(*Left*) *Phyllostachys viridis* lining an avenue in Prafrance

(*Right*) Culms of native *Dendrocalamus asper* reach 30 cm (12 in.) in diameter

There are large bamboo hedges there, and many bamboo species, some of which will only grow in the mild climate of southern France. A large proportion of the container plants sold in Germany today also come from Prafrance.

This remarkable bamboo collection has an interesting history. At the end of the last century Eugene Mazel, a prosperous French merchant, travelled to China to study silk production. He was fascinated by bamboos and brought some plants back on his ship. He planted these at Cap Ferrat on the Côte d'Azur and was amazed to see them grow and spread. Later on he bought 40 hectares of land near Anduze and dammed the river Gardon to supply his bamboos with water. Forty gardeners worked in the expanding bamboo garden for 20 years until Mazel got into financial difficulties and lost his ownership. The land was then leased to local farmers whose cattle eagerly devoured the bamboo shoots. It was not until the turn of the century that the garden found a bamboo enthusiast again, in the shape of Gaston Nègre, who bought the area and redeveloped the bamboo collection. It took two generations to bring it to its present state. Gaston Nègre's niece, Muriel Crouzet and her husband Yves have created from it a veritable Mecca for bamboo enthusiasts.

In the USA there are many established bamboo collections, in both private and public gardens. The favoured climates of the Southern and West Coast states enable tropical bamboos to flourish outside. At the Quail Botanic Garden in California, an important collection is being established by the American Bamboo Society and the plantings include many of the rarer and recently introduced bamboos.

Gardeners in Switzerland (particularly in Tessin), Italy and in other southern regions of France have also been successful, and have invested much effort and knowledge in their bamboo gardens. Even in Germany those in the know can find big bamboo stands, though these are very small compared with the huge groves of their Asiatic homeland. For example, in the new Hamburg botanic garden there is a very formative bamboo garden with very many genera and species, some of a most imposing size. Anyone can also visit the Ellerhoop Arboretum between Pinneberg and Elmshorn, which has many bamboo species. Helgoland Island, with its mild climate, has some very decorative but frost-sensitive species, for example the Tortoiseshell Bamboo, *Phyllostachys heterocycla*.

3

Morphology and Structure

Botanical classification

Bamboos are a subfamily of the grasses (family *Gramineae*). For example, the classification of *Phyllostachys nigra* 'Henonis' is as follows:

Family	*Gramineae* (alternatively, *Poaceae*)
Subfamily	*Bambusoideae*
Genus	*Phyllostachys*
Species	*nigra*
Cultivar	'Henonis'

In short, bamboos are tree-like or shrubby grasses with woody stems. According to the latest classification there are 115 bamboo genera and between 1,020 and 1,070 species worldwide, although bamboo researchers and taxonomists are still not in full agreement over assigning certain species to particular genera. The precise identification of bamboos is difficult because they flower so rarely, sometimes at intervals of 120 years. Systematics however depends upon flower structure. Yet even with flowers the numerous *Sasa* species are difficult to distinguish.

Bamboos do not always grow in a characteristic way, since the growth form is highly dependent upon the particular site. A species growing in a protected site with a good microclimate can look completely different from a specimen of the same species growing in an unsuitable place with poor soil conditions and too much exposure. Sometimes the reasons for the differences are not at all clear. For example in Asia the Tortoiseshell bamboo, *Phyllostachys heterocycla* f. *heterocycla* 'Kiko', has thickened and distorted internodes, rather like a tortoiseshell, but in temperate climates it seldom produces these distortions. The reasons for the development of swollen internodes are not known.

Many American, Japanese and English authors have tried to produce a precise botanical classi-

fication for all known bamboo species, but none has proved to be definitive. Scientists cannot agree about the taxonomy. As has been pointed out several times, anyone who works seriously with bamboos will sooner or later encounter problems of identification because the self-same bamboo species often has two or even more different names. This is particularly acute in the case of the genera *Arundinaria* and *Pleioblastus*. For instance a plant going under the name *Arundinaria* in the Hamburg botanic garden is called *Pleioblastus* in Gruga Park (Essen) and *Thamnocalamus* somewhere else. This is because bamboo nomenclature is still in a state of confusion. Now that communications with China have improved there have had to be some changes to the classification. For example, the most widespread bamboo in Germany, *Sinarundinaria*, was first described in China as *Fargesia* and is also correctly named as *Fargesia*, even though the name *Sinarundinaria* is common and widely used in gardens and by suppliers. However, in this book the synonyms are quoted and can be found in the index. The bamboo that was named *Thamnocalamus spathaceus* just a few years ago is found here under *Sinarundinaria*.

Dieter Ohrnberger and Josef Goerrings spent many years dealing with bamboo nomenclature, classification and distribution in their work *The Bamboos of the World*. This is a bibliographical encyclopedia dealing with all species and their world distribution, evaluating all relevant literature from Linnaeus to the present day. Most taxonomists follow this basic classification, particularly the definition of the genera, although their own investigations are not yet completed. *The Bamboos of the World* is published in individual volumes in English and is very valuable, if not essential, for any scientific work on bamboos. For many gardeners, however, who simply wish to have one or a few bamboo plants in their garden, such precise scientific identification is not so important. However, it is still interesting to have some knowledge of their morphology and physiology, and indeed, necessary for the correct choice and cultivation of bamboos.

Morphology

Bamboo is an evergreen plant, that is it does not lose its leaves in the autumn and grow fresh ones in the spring like other broad-leaved trees. Bamboo leaves stay green throughout an average winter. In early spring the young leaves grow out and the old ones are gradually lost. This characteristic of bamboo makes it particularly desirable for our gardens. The soft-green foliage decorates a garden even when it is freezing and bamboo looks particularly good in the snow.

All in all bamboo is a very hardy and vigorous plant. Even when the stems and leaves have been severely damaged the plant will usually recover, although it may take years to regain its previous height. After the destruction of Hiroshima by atomic weapons it was the green bamboo stems that were the first sign of new life.

Bamboos have another characteristic that distinguishes them from the European native trees and shrubs. The young shoot that emerges from the ground has the diameter at which the bamboo culm will remain throughout its life, which may be ten years or more. Bamboos do not therefore grow in width like trees, which get visibly broader each year. The length that a shoot reaches in its growth year is also its final length – there is no year on year increase. Young developing bamboo plants produce thicker and taller stems each year, until the clump reaches maturity. The number of branches, however, does increase each year, as does the number of leaves.

The main structural parts of a bamboo plant are the underground system of rhizomes, the aerial culms and the culm branches. All of these parts are formed according to the same principle; an alternating series of nodes and internodes.

As a bamboo grows, each new internode is wrapped in a protective sheath, attached to the preceding node at the sheath ring. Once the internode has lengthened, it does not grow any further. The nodes are massive pieces of tissue, comprising the node ring, the sheath ring and usually a dormant bud. These dormant buds are the site of emergence of new segmented growth.

Rhizomes

Rhizomes are underground stems which grow and branch away from the bamboo plant, thus enabling new territory to be colonized. Each year,

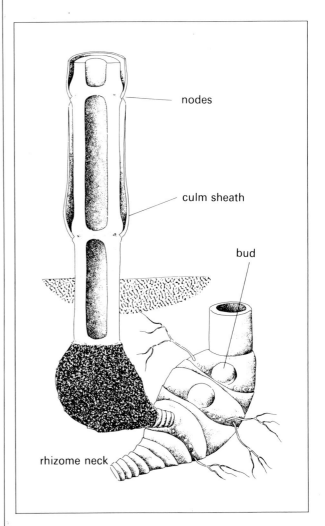

Culm of a clump-forming bamboo. The section shows that the culm is thinner than the rhizome from which it grows. The culm sheaths develop from the sheath ring at the nodes which divide the culm into segments

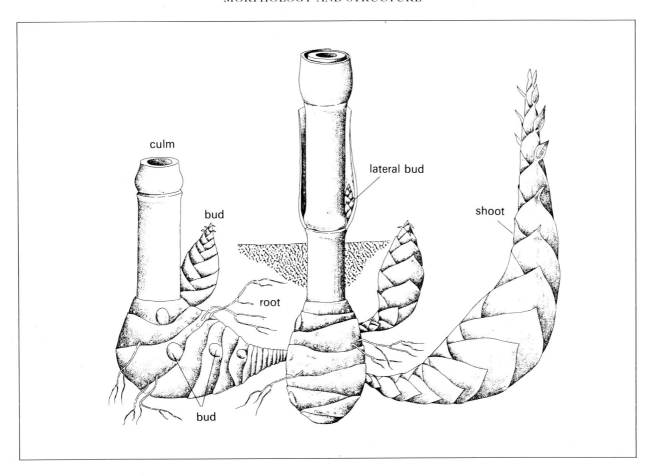

culm

bud

lateral bud

shoot

root

bud

Pachymorphic rhizome. Bamboos with pachymorphic rhizomes have a clumped growth. The short, thick, often sickle-shaped, rhizomes grow upwards to form a culm. At the nodes are the lateral buds from which fresh rhizome develops, connected to the old rhizome by a short stem, known as a rhizome-neck

culms arise from the rhizomes to form the aerial parts of the bamboo. These rhizomes are often so tightly packed that the soil under a bamboo plant seems to be filled with them. They form a 'turf' similar to ordinary grasses, which can vary in depth, depending on the species and growing conditions, although seldom deeper than one metre (three feet).

In Japan, people are said to flee into bamboo groves during earthquakes, because the soil is bound together with unbreakable rhizomes!

As with all segmented parts of a bamboo, growth takes place only at the tip of a rhizome, which is the site of cell division. The rhizomes grow on, continually branching, producing new plant tissue which differentiates alternately as node or internode. The growing point of a rhizome is very hard, allowing it to penetrate into the surrounding soil. It is formed of tightly overlapping sheaths, which enclose each developing internode as the rhizome moves forward. These sheaths are very short-lived, quickly dying and rotting away once the new internodes have hardened.

The true roots of a bamboo develop from the nodes of a rhizome, to supply the bamboo with water and nutrients from the surrounding soil. They emerge from the node ring and are usually thinner than the rhizome and are not segmented.

Rhizomes can be broadly divided into two growth types; pachymorph and leptomorph. The habit of the bamboo grove above ground is dependent upon the rhizome type, as this deter-

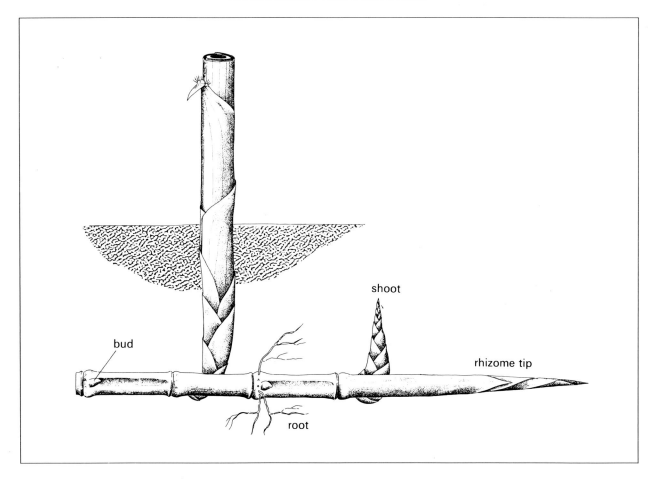

Leptomorph rhizome. This grows horizontally and is long and noticeably thinner than the aerial culms. The buds give rise either to a culm or to a fresh rhizome

mines whether the grove will be invasive or clump forming and will also determine the distances between individual culms.

Pachymorph rhizomes

This type of rhizome is most usually encountered in tropical bamboos, such as those of the genus *Bambusa*, although also occurring in some temperate bamboos. The rhizomes are short and thick, usually curved upwards, and solid. The internodes are very short and look compressed. From the dormant buds on the nodes of a pachymorph rhizome, only new rhizomes can develop. A culm is produced from the tip of each rhizome as it

curves upwards, and is usually thinner in diameter than the rhizome from which it has developed. Above ground, the culms of this type of bamboo are usually very close together in a tight clump formation, which expands evenly round its circumference. This type of clump habit is also known as sympodial and such bamboos remain in their original site and expand laterally only by very short distances each year.

Leptomorph rhizomes

Bamboos with leptomorph rhizomes are usually from the temperate regions and include the genera *Phyllostachys* and *Pleioblastus*. These rhizomes are long and thin, growing horizontally over considerable distances. In contrast to pachymorph rhizomes, either new rhizomes or culms can develop from the dormant buds on the nodes. When the rhizome has completed its seasons growth, sometimes the tip will grow upwards to become a culm, but the majority of the culms will be produced alternately on the left and right sides of the rhizome. The culms are thicker in diameter than the rhizome from which they have developed. Bamboos with this rhizome type, produce a more open clump habit, known also as monpodial, where the culms have a much greater spacing, often in straight lines away from the mother clump. Such bamboos are invasive and care should be taken when planting in the garden to allow room for future growth.

Intermediate forms

Not all bamboos fit exactly into the two main rhizome categories. Growing conditions and the general health of a bamboo plant can have a marked influence on vigour and clump habit.

Bamboo rhizome, exposed to show growth

In cool temperate climates, many of the taller leptomorph bamboos will be much less invasive and may grow in tighter clumps. However, some bamboos are naturally intermediate in their rhizome type. For example, *Arundinaria anceps* is basically a pachymorph bamboo but has an invasive habit. This is due to each rhizome having an elongated 'rhizome neck', that is, the length of stem between the rhizome proper and the parent rhizome from which it has developed. Such intermediate rhizome forms are common, particularly among bamboos from high altitudes.

Culms

Bamboo species are most readily distinguished by their culms, whose form and appearance give the whole plant its character. In some species the culms grows to several metres, in others just a few centimetres. Some species have thick culms, other thin; some grow erect, other bend graciously. Bamboo culms are normally green, but some are yellow, brown, black, reddish, spotted or striped.

The typical bamboo culm, as pointed out above, is divided into many segments comprising nodes and internodes. The culm is stabilized by the

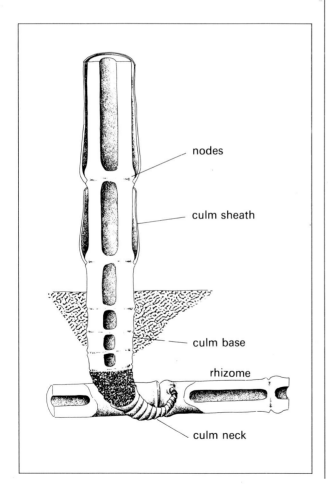

nodes

culm sheath

culm base

rhizome

culm neck

(*Left*) Culm of a leptomorph bamboo. It develops from a bud at a rhizome node. The culm is always noticeably thicker than the rhizome and, as the drawing shows, culm and rhizome have the same overall structure

tough nodes, as is the case in grasses and cereals. In most bamboo species the nodes and internodes alternate evenly, but there are exceptions. For example, *Phyllostachys aurea* often has slanting, irregular nodes and shortened internodes in the lower part of the culm. The 'tortoiseshell' bamboo (*P. heterocycla* f. *heterocycla* 'Kiko'), which grows in Japan has nodes that are not horizontal, but are set at an angle of 45 degrees in the lower part of the culm, giving the appearance of a small tortoiseshell. Most bamboo culms are round, but those of *Chimonobambusa quadrangularis* are often markedly square in cross-section, particularly near the culm base. This species also has small aerial roots on the basal nodes.

Most bamboo species have hollow internodes. However here there are exceptions, for instance *Chusquea*, from Central and South America, whose internodes are solid. In Asia the culm is the most important part of the bamboo plant, since so many objects used in daily life derive from it. One therefore talks of bamboo wood, even though in the strict sense bamboo is not wood. Bamboo does consist of cellulose fibres, but whereas in tree wood these fibres are usually only about a millimetre long, in bamboo they are up to a centimetre. These long fibres contain lignin and silica, whereas tree fibres only have lignin. The proportion of silica in bamboo can be measured by burning the stems and measuring the silica in the ash. Bamboo contains up to five per cent, with young stems containing the most.

Culm growth

In summer and autumn the rhizome builds up and stores all the material that is required for rapid growth the following spring. These reserves are mostly the raw materials for tissue growth produced through photosynthesis by the plant. The shoots, which are usually edible, begin to develop underground in the autumn and contain all the basic organs of the culm.

The shoots appear above the soil at a specific time of year, depending on the species, and indicate the diameters of the culms which will develop from them. They may be as thin as grass

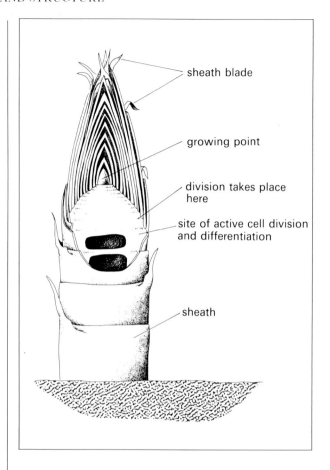

sheath blade

growing point

division takes place here

site of active cell division and differentiation

sheath

The bamboo shoot emerges at its final thickness with the internodes already laid down in the shoot. These grow out from each other in a telescopic fashion, protected by the culm sheaths. They also protect the growing point, which creates new material by cell division and differentiation. The small 'leaves' at the tip are the sheath blades

or as thick as 25 cm (10 in.) in diameter, according to the species or age of the plant. The eventual number of internodes is pre-determined in the shoot, which grows out telescopically and with great speed. When the shoot emerges from the soil it appears as a series of tightly overlapping sheaths surrounding the developing nodes and internodes. These have the same function as those of the rhizome; to protect the soft culm as it grows, and to provide growth hormones. If they are removed too soon, then growth of the internodes ceases; bamboos can be cultivated as bonsai by this method (p. 111). If a bamboo shoot is cut off before it has begun to elongate, for example by

Attractive bamboo culms. From left to right: *Phyllostachys aurea, Chimonobambusa quadrangularis, P. nigra, P. bambusoides* 'Violascens', *P. nigra* f. *boryana, P. viridis* 'Robert Young', *P. bambusoides*

mowing, the culm dies. If it is cut after the culm has started to develop, it will branch out and produce foliage.

The bamboo shoot grows quickly, especially in the first week. It is interesting to place a rule next to a growing shoot and measure an astonishing growth rate of several centimetres a day. Large species in good sites can grow as much as 40 cm (16 in.) in a day.

The length of internode appearing from the culm sheath towards the end of the growth period varies with the species and is longer in those species in which the growth period begins earlier. When the elongation period is over the culm sheaths dry up and fall off. These sheaths are sometimes attractively coloured (in certain *Phyllostachys* species they are reddish and shiny, in others more or less hairy), providing an attractive contrast to the rest of the plant. In most genera they fall off straight away but in some genera, such as *Pseudosasa* and *Sasa*, they may persist for two years, or even longer.

The culm sheath consists of the sheath and blade. The blade is a small leaf-like protrusion from the tip of the sheath. Between sheath and blade, many species have auricles (ear-like projections) with bristles, sometimes with a fringed ligule as well (see diagram below). Some species lack auricles altogether. The blades of the sheath increase in size towards the tip of the culm and may develop into proper leaves. In many *Phyllostachys* species the blade is completely developed inside the shoot. This means that they are folded up, occupying just a tiny space. Such blades, when they appear, often have not only pretty colours but also very interesting patterns. Some are creased, others pleated or wavy.

Phyllostachys culms have a noticeable groove, the sulcus, running along the internodes above the side branches. In *Semiarundinaria* this is only partially present, particularly in the lower internodes. These grooves exist because of the way the branch buds are formed in the developing shoot.

Some bamboos, like *Sinarundinaria nitida* and *Chimonobambusa*, produce only culm in one growth period, but no branches or leaves until the next year, giving a characteristic appearance. The naked pointed culms stretch upwards with only the small sheath blades attached, in an otherwise leafy bamboo clump. In *Sinarundinaria nitida* these naked culms are more or less hardy. In cool summers *Sinarundinaria murielae* develops only leafless culms and these appear from the soil rather later than usual. They are 'unripe' however, and can scarcely withstand frost. *Chimonobambusa marmorea* and *C. quadrangularis* can even grow fresh culms in September, depending on the climate. If such culms survive the winter they develop their branches and leaves the following spring.

(*Below*) Culm with sheath and blade. The culm sheath develops from the sheath ring at the nodes. During the first few days, when the growing internode is still young and soft, the culm sheaths protect it. When internode growth ceases the culm sheaths dry and often take on attractive colours. They stay attached for a variable time, depending on the species. At the tip of the sheath is a small blade, and the higher up the culm these are situated the longer they are, until on the branches each blade is large enough to be called a 'leaf'. Between sheath and blade is the ligule with a fringe, auricles and bristles. In some species these are very obvious, in others almost invisible

(*Right*) *Phyllostachys heterocycla* f. *pubescens* in the grounds of a temple in Kyoto, Japan

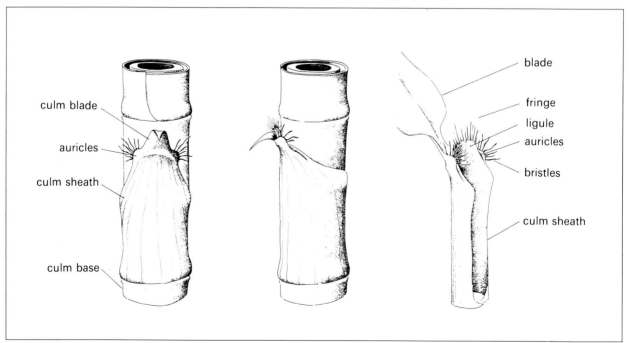

Phyllostachys heterocycla f. *heterocycla* 'Kiko', the 'Tortoiseshell' bamboo

Branches

As the bamboo culm grows, protected by its sheath, branch buds develop at the nodes. In *Phyllostachys* and *Semiarundinaria* the bud is laid down in the shoot and, therefore, in these genera the branches appear as the culm elongates. In most other genera the branches start to appear only after the main culm has completed its elongation. In some, such as *Sinarundinaria* or *Chimonobambusa marmorea*, branches do not develop for two or three months, or even not until the following spring. In some genera the branches start developing from the top downwards along the culm, in other genera it is the other way round. Certain genera, such as *Sasa*, *Sasaella* and *Pseudosasa* only grow branches on the upper culm, whilst other develop branches all the way along. However this also depends to some extent on whether the plant has sufficient light during the period of culm growth.

The number of branches at the node is an important taxonomic characteristic in identifying bamboos, although there is some variation. *Sasa* species have only one branch per node, and this is true also for *Pseudosasa* and *Sasaella* (which may occasionally have two to three branches). *Phyllostachys* usually has two, a stronger branch and a smaller one about two-thirds the length. There is sometimes a third, even smaller branch between these two, which, on close inspection is seen to arise from a bud on one or other of the longer branches. In *Phyllostachys*, well-grown plants occasionally have only a single branch on some lower nodes. Lack of light weakens one

Phyllostachys aureosulcata

branch, which then withers and dies. *Pleioblastus*, *Semiarundinaria*, *Sinobambusa* and *Chimonobambusa* have three branches per node; *Arundinaria* and *Sinarundinania* three to six branches; *Dendrocalamus* and *Bambusa* seven to nine. *Sinarundinaria* has many thin branches and *Chusquea* has a whorl of up to 50 branches surrounding each node.

Close inspection of the nodes often reveals that not all the branches arise from buds at the base of the main internode, especially in those genera with many branches. Often some of them grow from buds on the first branches but because these have such compressed internodes it seems as if all the branches arise from the same bud.

The leaves develop on the branches and twigs, and so too do the buds for the following year's branches.

Phyllostachys nigra has striking black culms

Bambusa glaucescens 'Alphons Karr'

Leaves

A bamboo leaf does not simply grow out from the bud, as for example would the leaf of a broad-leaved tree such as beech. The bamboo leaf develops from a sheath, which encircles the stem and is called a leaf-sheath. The leaf-sheath resembles a small culm-sheath but develops a large sheath blade which functions as a proper leaf. The leaf blades (to give them their proper term) are attached to their sheaths by a short, stalk-like projection of their midrib, which, when the 'leaves' fall, breaks off from the sheaths, which remain attached for much longer.

At the point of attachment of the blade to the sheath there are different combinations of features involving auricle, bristles and ligule, which are as

Culm of *Phyllostachys* with sulcus, a groove running the length of the internode above the side branches. In *Semiarundinaria* the sulcus is only present on the lower half of the internodes. In these genera the bud from which the branches develop is already present in the shoot and makes a groove in the soft internode as the culm extends

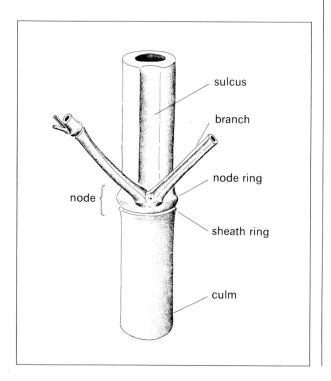

important as the colour of the culm sheath in the identification of individual species. The bristles prevent water from entering the leaf-sheath and one often sees drew drops glistening on them. The presence of these features varies from species to species, but can vary quite a lot even within a species, according to the growing conditions.

A bamboo leaf is elongated and lanceolate, rounded at the base and pointed at the tip. Length and breadth vary markedly between the different genera and species. It is a characteristic of bamboos that the 'leaf' (actually the leaf blade) always has a stalk, unlike the leaves of other grasses. In most grasses the base of the leaf is directly wrapped around the stem, without a stalk. Bamboo leaves have stalks for a good reason. They are evergreen and the leaves must be elastic so that they can survive winter snow and rain. Without a stalk such elasticity would be difficult. In some species

Branching patterns.
The number of branches is an important characteristic in distinguishing bamboo genera. *Sasa* has only one branch per node, *Phyllostachys* has two main branches, with sometimes a third, weaker branch growing from a bud on one of the others. *Sinarundinaria* and *Arundinaria* have three to six branches

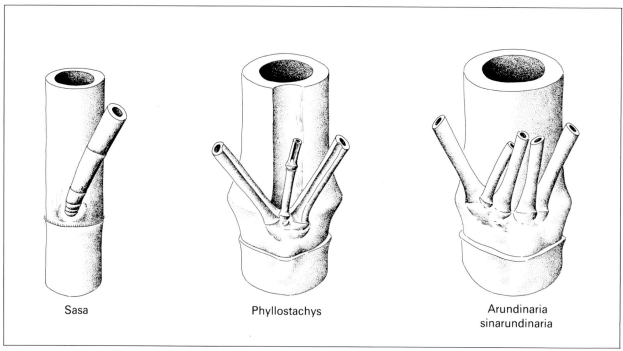

Sasa

Phyllostachys

Arundinaria
sinarundinaria

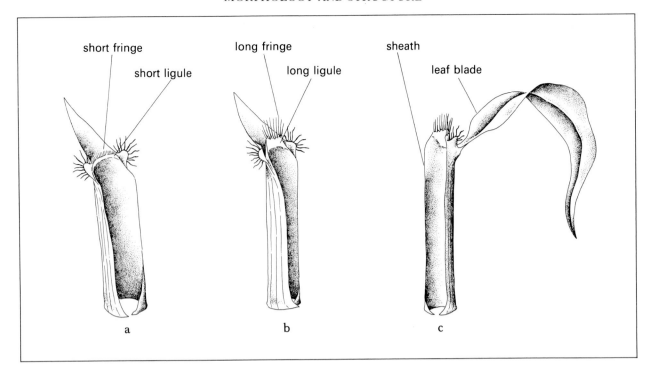

Bristles and fringes.
Between the sheath and blade is the ligule with its fringe. At either side are the auricles, with bristles. The fringed ligule and the bristles prevent rainwater from flowing down inside the culm sheath. All these features vary in length from species to species. The culm sheaths (*a* and *b*) have relatively short blades, but these are fully developed as leaves on the branches (*c*)

the stalk is only a few millimetres long, in others much longer. Bamboo leaves can live as long as two years in certain species and this would not be possible without a stalk. The stalk allows the leaves to move and be adaptable, and this movement also makes them particularly attractive. The leaves move at the slightest breeze and continually create new sounds, shadows and rustles.

Each branch continues to grow new leaves throughout the summer. However, bamboos do lose some leaves in the autumn and produce more in spring. *Phyllostachys* and *Sinarundinaria*, for example, lose several leaves in the autumn, whereas in other genera, like *Sasa* and *Pleioblastus*, the leaves do not fall but wither around the edges and the tip, thus reducing the area of transpiration in the winter.

Unlike European and American evergreens, like ivy or holly, bamboos have extremely thin

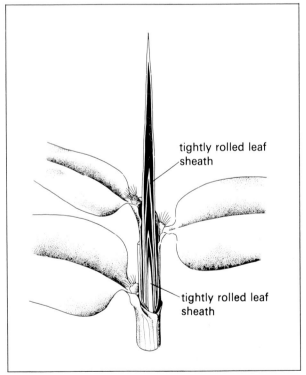

Bamboo leaf. On the branches the blades are developed as leaves. The leaf sheaths are tightly rolled round the thin branches, obscuring the internodes and nodes. The leaves may have short or long stalks, depending on the species

leaves. One would therefore expect them to be particularly frost-sensitive. However, they have a special adaptation that helps them. Bamboo leaves have parallel 'veins', like other grass leaves. These leaf veins run parallel over the whole length of the leaf and are connected other veins running at right angles, to give the characteristic pattern known as tessellation. In some species this is easy to see, but in others you need a magnifying glass. Visible tessellation, that is, thick cross 'veins', is usually a sign of winter hardiness. In cold weather the volume of cell sap goes down and the leaf cells develop a low pressure. The strong cross 'veins' prevent cell destruction and consequent leaf damage. Examples of genera with strongly tessellated leaves are *Phyllostachys*, and *Pleioblastus*.

Bamboo leaves vary a great deal in colour and shape, and these can be important for identification. The choice of a bamboo species for the

(*Left*) Foliage of *Phyllostachys hetrocycla* f. *pubescens* (seedling stage)

(*Below*) Development of leaves and branches. The rolled-up leaf blades open out from the telescopically arranged leaf sheaths. As the branch grows, the next leaf sheath appears, rolled round a branch internode

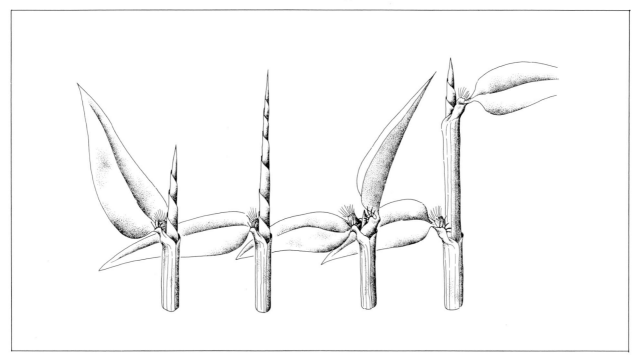

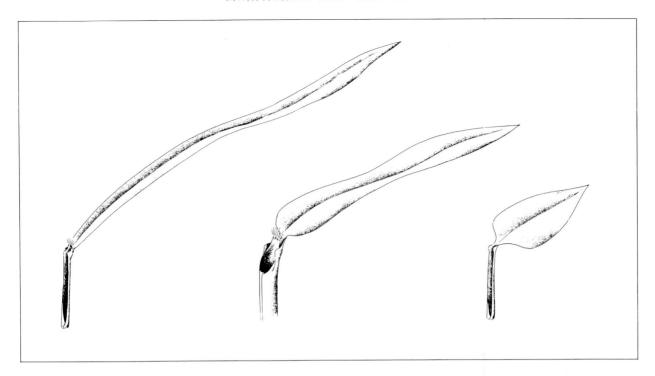

(*Above*) Leaf shapes. The shape is characteristic of each particular genus. *Sinarundinaria* species have very narrow, long leaves, *Phyllostachys* somewhat broader, whilst in *Sasa* species the leaves are usually sturdy, and are often very broad

(*Right*) Young bamboo shoot with sheath blades

garden will certainly depend upon whether the leaves are delicate or broad, whether they grow densely or more open. Certain *Pleioblastus* species, for example, have very long, narrow leaves, as does *Otatea acuminata* ssp. *aztecorum* (this is not, however, suitable for growing outdoor). All *Phyllostachys* species produce comparatively medium-sized leaves, and in *Sinarundiniaria* the leaves are somewhat smaller and rather narrow. Most species of *Sasa* and *Pseudosasa* have relatively large, wide leaves, and *Indocalamus tessellatus* has the largest leaves of any hardy bamboo, up to 60 cm (24 in.) long and 10 cm (4 in.) broad. In any case, the size of bamboo leaves can vary, depending on growing conditions and the health of the plant. The young culms of a well-grown plant will usually have larger leaves than that of a similar plant growing in an unsuitable site. In some species the underside of the leaf is darker or even blueish. Species

41

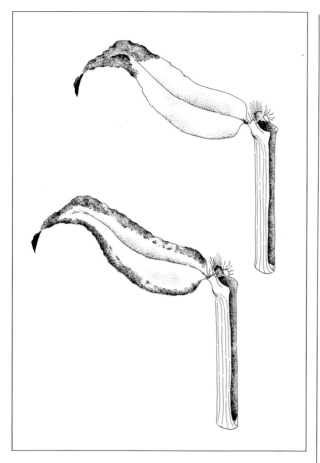

Most bamboos lose and renew their leaves on a continual basis. During this process, the leaf dries from the tip, then at the edges, before falling off altogether

with green-and-white or green-and-yellow striped leaves are quite sought after, for example *Pleioblastus variegatus* or *P. argentiostriatus*. Another kind of colour develops when bamboo leaves dry off from the tip or sides of the leaf blade, losing their chlorophyll. The best example of this is the low-growing species *Sasa veitchii*, in which the leaves develop white withered edges in winter.

The bamboo flower

Bamboos do not usually flower at regular, yearly intervals, like the majority of trees and shrubs, but come into flower only very rarely. The flowers are very inconspicuous and in some instances (though not as a matter of course), the plant may die after flowering. These are three fascinating characteristics of bamboo, but it is the flower itself that most fascinates botanists and laymen alike. This is because the bamboo flower holds many mysteries that have still not been fully explained, although many theories abound. What has been established is the most genera only flower at long intervals, of 10, 50, or even more than 100 years. For such long intervals some sort of explanation is clearly required.

This rarity of flowering makes the identification of individual species, and even their allocation to a particular genus, especially difficult. Botanical nomenclature depends upon flower structure, so one can see that studying and identifying a bamboo flower could be the work of a generation or more. For precise identification one should really collect seed, raise plants from this and study the resulting flowers. But how can this be done when a genus only flowers once in a hundred years? In 1912, when *Phyllostachys viridis* followed in Japan, scientists from Tokyo and Kyoto universities planted seed in two different research plots. The scientists who set this up are now long dead, the bamboos have not flowered again since, and no-one knows when they will do so. It is possible that the flowers will first be seen by the scientists of the next generation but one.

For a long time it was thought (and this can be traced in the old scientific literature) that all plants of a particular species flowered simultaneously all over the world, in Asia, America and Europe for example. However, this is not strictly true. For example, in 1982 *Pseudosasa japonica* came into flower across Europe. In one garden certain clumps flowered whilst others did not and these were probably different clones. Some species of the South American genus *Chusquea* recently flowered repeatedly in its natural habitat, but only a few weakly flowering plants have been noticed in Europe and America. In general one can say that when they do flower, bamboos in cultivation flower at roughly the same period, if they are of the same clone. For example, *Phyllostachys bambusoides* 'Castillonis' flowered in 1963 in England, but other clones elsewhere in

Europe did not flower until 1977. Fifteen years passed between the onset of flowering and the time when the last bamboo began to flower. There are possibly even greater discrepancies, but bamboo research in Europe is still in its infancy.

This simultaneous flowering (assuming flowering occurs at all) is certainly connected with the

Anther of *Pseudosasa japonica*

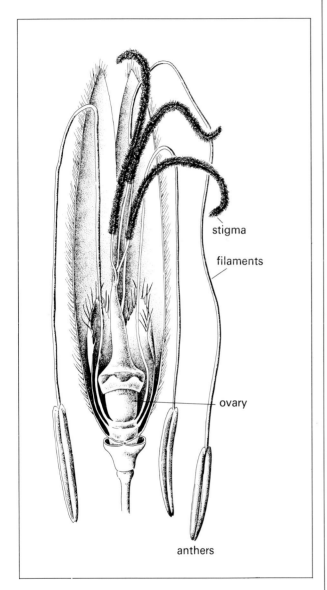

The bamboo flower is inconspicuous and, as in all grasses and cereals, ear-like. The drawing shows schematically a flower of *Phyllostachys*, much enlarged. Wind carries the pollen from the bright yellow anthers to the stigmas, the long filaments assisting this process. After pollination the cereal-like grain develops from the ovary. This, the seed of the bamboo, remains viable only for a very short period

method by which most bamboos are propagated. Bamboos, whether wild or in cultivation, are propagated by division of the rhizome (see pp. 92–6). In addition, new species are introduced not as seedlings, but as divisions. Most bamboo plants are therefore clones, derived from a single plant, with the same genotype. This is probably why they flower simultaneously.

There is a story relevant to this subject. A few years ago *Sinarundinaria murielae* had not yet flowered in cultivation. Then, in 1984, plants of this species began to flower in Denmark. Thomas Soderstrom, of the botany department of the Smithsonian Research Institute in Washington, investigated the flowers and identified the plant as *Thamnocalamus spathaceus*, on the basis of flower structure. At that point the matter became complicated. Should all *Sinarundinaria murielae* be renamed? Or was the Danish material a different species? But *Sinarundinaria murielae* did not flower in Germany, France or Italy. Seedlings of the Danish plants were raised and propagated and soon fetched a high price under the name 'Danish clone' of *Thamnocalamus spathaceus*.

Many bamboo species – but not all – die after flowering. The reason is that flowering uses up all the plant's energy. Few fresh leaves are produced during flowering so that there is not sufficient leaf surface for photosynthesis.

Flowers of *Phyllostachys elegans*

When a bamboo starts flowering it produces new flowers instead of leaves. At first there are just a few flowers, then more and more. Some develop a few small leaves near the flowers, for example *Pseudosasa japonica*. Others develop leaf-like bracts at the tip of the flower. Even new growth in the spring can be pure flowering growth. In this way, some genera, like *Sasa*, *Sasaella* and *Pleioblastus* can even have a kind of interval within flowering. They produce normal stems with leaves in the spring and then start flowering in summer. Other genera give themselves over completely to flowering, producing thousands of flowers over several years. Such a plant relies almost entirely on food from the rhizome. This uses up a lot of reserves, since, as with every flowering plant, it takes more energy than making leaves or fresh growth. In each growth season the stems get progressively smaller and thinner, but continue to flower. Such plants may eventually die, or get so weak that it takes them years to recover. It

is impossible to help such plants by fertilizer application or heavy watering, or indeed in any other way. One simply has to accept the consequences of bamboo flowering in the garden as a natural phenomenon. If you are lucky you will be able to collect seed, or the bamboos may have sown themselves, to produce seedings. One should try to plant bamboo seeds in a light compost, under glass, and you may be lucky and grow a seedling which can then be planted out p. 83.

Flowering *Pseudosasa japonica*

4

Characteristics of Bamboo Genera

This is a controversial subject, open to much disagreement and criticism. Plants are classified according to their flower but because bamboos flower so rarely, many bamboos have been classified according to their vegetative characteristics. This has led to much name changing in the past, where the main botanical authorities have been unable to agree between themselves the correct classification of certain species. Whenever a bamboo has finally come into flower, this has frequently resulted in a name change. The genus *Arundinaria* is perhaps the most controversial. It was once a very large genus with numerous species but has since been divided into several smaller genera, with considerable disagreement among the botanists and taxonomists.

Arundinaria

Arundinaria Michaux is a genus with an indeterminate number of species. It is native to the Himalayas, but also China, America and S. Africa. Many *Arundinaria* species have attractive and persistent culm-sheaths, which give the plant a distinctive appearance. In some *Arundinaria* species the young shoots are coloured. This genus characteristically has three to six branches per node. Some species only develop these in the second or third year, with none at all in the first year. The leaves are not particularly characteristic, and vary a lot from species to species – long and wide in *A. gigantea*, but short and narrow in others. In some species, for example *A. gigantea* ssp. *tecta*, the leaves are softly hairy on both sides. The stems are round and 1.5–8 m (5–26 ft) tall in good conditions. *Arundinaria* species are mostly hardy, except in areas with a cold, continental climate, where they are best suited to containers or the conservatory. In England *Arundinaria* is often grown as a hedge plant but in central Europe the winters are mostly too cold (except in the wine growing areas) for successful cultivation in the open. Most *Arundinaria* species like shade or semi-shade. The thin bamboo sticks that one sees for sale as supports for garden plants are mostly from *Arundinaria amabilis*. These can be distinguished from the thicker *Phyllostachys* species because they do not have a sulcus.

Bambusa

The genus *Bambusa* Schrader, from tropical Asia, America and Africa, has more than 100 species. They are only suitable for growing in containers, which can be brought under glass in the winter, for warm conservatories, and for use as house plants. *Bambusa* always has several long branches at the nodes. The culms are mostly green, but not in all species. *B. vulgaris* 'Vittata' has yellow culms with green stripes and is therefore particularly valued as a decorative plant, both here and in its country of origin (S. China). All *Bambusa* species cultivated in temperate climates produce their culms in late summer, rather than in early spring and if the base is big enough, these can grow as high as 15 m (40 ft).

A speciality of the genus is *B. ventricosa* (Buddha's Belly). Although the author has not tried this method, if it is planted in a pot that would normally be regarded as too small, watered infrequently and not given too much fertilizer, then the above-ground internodes on the new growth thicken up and turn into veritable 'Buddha bellies'. In a larger container the culm grows upright and slim in the normal manner. Another speciality is *B. glaucescens* 'Fernleaf'. It has usually

solid stems and each individual branch has up to 20 leaves in two rows, making it look like a fern. All *Bambusa* species are strongly clump-forming.

Bashania

In the late 1980s the Chinese genus *Bashania* P.C. Keng and T.P.Y. was introduced to Germany, with a single species. *B. fargesii* resembles *Pseudosasa japonica*, and both have leaves 20 5 cm (8 10 in.) long and up to 5 cm (2 in.) broad. This species grows to about 3 4 m (10 13 ft), puts out long rhizomes and is frost-hardy. It is often offered under the name *Arundinaria fargesii*.

Brachystachyum

The genus *Brachystachyum* Y.L. Keng comes from China and is not well known in the West. It throws out long rhizomes and withstands most winters very well. In Europe and America they do not grow taller than 3 4 m (10 13 ft), with leaves up to 20 cm (8 in.) long and 2 4 cm ($\frac{4}{5}$ 1$\frac{1}{2}$ in.) wide. *B. densiflorum* is also offered under the names *Arundinaria densiflora* or *Bashania densiflora*.

Chimonobambusa

Chimonobambusa Makino, a genus with several species, from the Himalayas, China and Japan, is best suited as a container or house plant, except in mild climates. In summer *Chimonobambusa* species are particularly good for courtyards or balconies, since they tolerate semi-shade. The three species available in Europe and America grow to about 3 m (10 ft) high in containers and have a particularly attractive growth form. One cultivar, *C. marmorea* 'Variegata' has white-striped leaves.

(Left) Bambusa vulgaris grows very large in its native China. Elsewhere it must be kept as a container plant, or in the conservatory, and therefore remains much smaller

The plants send out new culms in late autumn and develop more and more twigs at each node in early spring. The culm-sheaths are light with brownish marbling and fall off within a year. *C. marmorea* has a small ring of hairs at the base of the leaf sheath.

Chimonobambusa develops long rhizomes when planted out. This is of no particular value, however, since in temperate climates the plants must often be grown under the protection of a greenhouse, in a conservatory or as house plants.

C. quadrangularis is often grown in tropical houses in botanic gardens. It is worth a mention here, although it is very difficult for the amateur to cultivate since it easily grows much too tall under glass and is too tender to grow outside. Its culm is markedly quadrangular with rounded corners. Under glass it can grow to up to 8 m (26 ft) tall and is therefore suitable for foyers and large rooms, but should not be allowed to stand in the full heat of the sun. The height and thickness of the culms depend on the size of the container and if you want an attractive large plant you have to use a big container. This species does not produce many new culms and, as in the tropical bamboo species, any new growth does not appear at the beginning of the season but towards the end, in autumn or even in winter.

Chusquea

The genus *Chusquea* comes from S. America where there are about 90 species from Mexico to S. Argentina. We see this genus only very rarely. In the trade two species are available, *C. couleou* and *C. coronalis*, but the latter is rather frost-sensitive and very difficult to cultivate. *C. couleou* needs an oceanic climate and *C. coronalis* is tropical. *Chusquea* has very fine, soft leaves, mostly on branches from the nodes. *C. coronalis* develops a ring of fine branches around the stem. The consequence of this is that a *Chusquea* plant does not grow like a typical bamboo, and for this reason they are particularly sought after. The stems of *Chusquea* species are completely solid.

Dendrocalamus

In its native tropical Asia *Dendrocalamus* Nees develops the tallest bamboo culms. There they reach well over 15 m (50 ft) in height and 30 cm (12 in.) thick. The 30-odd species of *Dendrocalamus* are strongly clump forming but because they grow so big they are hardly suitable for the conservatory or containers and you often see them in the tropical house in botanic gardens, where they can grow as big as the greenhouse roof allows.

Drepanostachyum

Drepanostachyum P. C. Keng is a genus that contains a large number of species that are still to be identified, or species previously assigned to other genera. At the time of publication, only two species are commonly available in Europe: the upright *D. hookerianum* and *D. falcatum* with soft, arching stems. Both species are clump formers and are primarily suitable as container plants.

Hibanobambusa

× *Hibanobambusa* (Maruyanma & H. Okamura) is only recently obtainable in the trade. However, it is cultivated and propagated by many enthusiastic amateurs. It is one of only a few genera known to be of hybrid origin.

In Prafrance the genus is regarded as hardy. Two hybrid forms are under investigation: × *H. tranquillans* 'Shiroshima', with long, and very pretty yellow-and-white striped leaves, and *H. tranquillans* 'Kimmei' with green leaves. Both are notable for the very long hairs on their leaf sheaths. They grow to about 1.5 m (5 ft).

Indocalamus

Indocalamus Nakai is a relatively hardy genus with about 20 species in China and Malaysia. In Europe and America it grows at most to hip height, in loose stands. The culms are thin and green and the branches single. The leaves are rather large and strong. *Indocalamus* does not produce particularly long rhizomes and is therefore suitable for ground cover between shrubs, particularly as it does well in semi-shade.

I. tesselatus has pretty, arched stems and is therefore a good container plant. *I. latifolius*, recently introduced from China, is a larger species, growing to 2–3 m ($6\frac{1}{2}$–10 ft).

Otatea

Otatea (McClure et E.W. Smith) Calderon et Soderstrom comes from Mexico and Guatemala. Only two species are well known and these have very soft, fine leaves. As a tropical genus it is suited to containers or the conservatory, but is difficult to grow even there. Certainly a plant for the enthusiast, which requires more light than a temperate climate can usually provide, even in a conservatory. *O. acuminata* ssp. *aztecorum*, which is the only commonly available species in Europe and America, can grow to 8 m (26 ft).

Phyllostachys

Phyllostachys Siebold et Zuccarini is the most widespread and varied genus. The home of this large genus (it has over 60 species) is China, but it is also found in Vietnam, India and Nepal. Its elegant shape explains why it features in the foreground of so many Asiatic bamboo paintings. The stems are usually straight, bending over at the tip in some species. When growing in a thick clump the leaves develop high on the culms. This means that the culms, which are striped or spotted in many species, are therefore clearly visible. *Phyllostachys* culms are unusual in that they have a sulcus. In young plants this sulcus is visible from the lowest internode right to the tip of the stem, but in older plants only on those internodes that carry branches. In *Phyllostachys* the branch buds appear in the shoot. When this grows the bud makes the notch in the stem (see page 36),

Phyllostachys viridiglaucescens by a stream

which hardens into the sulcus. *Phyllostachys* is often attractively coloured, with striped, black or spotted culms, and its loose, elegant growth makes these colours still more decorative. Some species have raised nodes, in others they may be set at an angle, producing irregular internodes. The 'tortoiseshell bamboo' takes its name from the nodes and internodes, which resemble tortoiseshells. There are also *Phyllostachys* species in which the lower parts of the stem do not grow straight but in a curve, or even a zigzag.

The culm sheaths fall off early and in some species they are coloured and the sheath blades pleated. All species normally have two branches per node, with occasionally a third, very much smaller branch between. The leaves are light green, in some white- or yellow-striped, and in most species 10–20 cm (4–8 in.) long and 1.5–

3 cm ($\frac{3}{5}$–1 in.) wide. Occasionally they are much larger at the ends of young culms.

Most species are very vigorous, growing quickly and powerfully under good conditions and recovering well from damage, but not from flowering. A few species are also cold-tolerant and efforts are being made to introduce even hardier species from China. Certain rare species such as *P. dulcis* or *P. vivax* are hardy and the latter grows strong stems relatively quickly, and is frost-tolerant.

Phyllostachys is very suitable as a solitary plant because of its beautiful shape. In good sites most species can reach 8 m (26 ft) or more. In warm, dry areas the stems grow thick and woody, straight and strong, whereas in an oceanic climate with relatively high humidity and cool summers, the same species develops thinner, more bendy stems.

Many European bamboo groves are *Phyllo-*

stachys. It takes a long time in such climates, however, for *Phyllostachys* to make a grove because the rhizomes do not grow so readily. However, this is also an advantage since it is easy to keep this genus confined to a particular area in the garden.

For the best growth *Phyllostachys* needs careful siting in the garden. In summer it likes full, even hot sun, and a plentiful supply of water. In winter, however, too much sun is damaging, so the ideal site is one in which there is full sun in summer and partial shade in winter, when the sun is lower, from a tree or the wall of a house.

Pleioblastus

The genus *Pleioblastus* Nakai includes about 20 species, native to China and Japan, and which are very diverse in growth and appearance. The genus is therefore divided into three sections. With the exception of *P. simonii* all European and American species are from the section 'Nezasa'. *Pleioblastus* species range from 40 cm–4 m (16 in.– 13 ft) tall. Some, like *P. chino* and *P. viridistriatus*, have long, broad leaves, others have very fine, small leaves. Some have hairy leaves, others smooth, some with white or yellow stripes. The leathery sheaths with distinct blades remain on the stem for a long time. All *Pleioblastus* species produce long, powerful rhizomes in good sites and a single plant therefore spreads quickly. It makes good ground cover under bushes and trees, since it thrives in semi-shade and it is also successful as cover in sunny spots. The strongly growing rhizomes help to keep weeds in check when *Pleioblastus* is used as ground cover. The comments on p. 89 concerning control of bamboo growth apply particularly to *Pleioblastus*.

Many medium-sized *Pleioblastus* species look good in pots or containers and they are also suitable for shady balconies and terraces. All the species available in Europe are from temperate zones and are therefore cold-tolerant.

(*Left*) *Pleioblastus viridistriatus*, with its yellow-and-green foliage, is very attractive as ground cover

Pseudosasa

Pseudosasa Nakai, with six species, comes from Asia. It is widespread, usually the one species *P. japonica* (also called Metake) and the variety *P. j.* var. *tsutsumiana*. Some English gardens often have a variegated variety: *Pseudosasa* is a bushy bamboo with broad, full, green leaves that grow thickly. The branches develop mostly at the end of the culm with usually just one side branch, and never more than three branches from one node. The straw-coloured culm sheaths characteristically remain hanging on the culm for months. The rhizome is leptomorphic, with new branches developing from side buds in the spring.

P. japonica has flowered repeatedly for the last five years or so in Europe. Up to now they have only flowered weakly and have produced leaves alongside the flowers. Flowering has weakened the plants but not completely damaged them since the rhizomes continue to be nourished during the critical period. *Pseudosasa* is suitable for planting on its own since it produces only single, albeit very long, rhizomes. It can also be used as a hedge plant. The garden forms of *Pseudosasa* with white- or yellow-striped leaves that one often sees in Japan are rarely available in Europe, although efforts are under way to import them.

Sasa

The genus *Sasa* is known as 'forest bamboo' because in its native Japan it is found mainly in montane forests, in very many species and varieties. Some species even grow above the tree-line. *Sasa* species do not grow very tall – in Europe and America between about 40 cm and 2 m (16 in. and 7 ft). Most produce just one branch at each node, which is thickened. The sheaths dry off and remain attached for a long time. Most species dislike direct sunlight and their leaves easily become damaged if too dry. In moist climates *Sasa* can be planted in full sun, otherwise one should choose partial or even full shade. Almost all species have an extensive underground system.

The rhizomes develop a complex network that, it is said, cannot even be destroyed by an earthquake. In Japan *Sasa* is planted to stabilize embankments because the rhizomes penetrate the soil well and bind it together. There are, however, disadvantages to the strong underground growth of *Sasa*. The plants need much water and nourishment to sustain such strong growth and a shortage can lead to an unattractive appearance. *Sasa* groves intended as ground cover should therefore be occasionally cut back to ground level.

Sasaella

Sasaella Makino has about 12 species and comes from Japan, and this genus is very similar to *Sasa*. All species have erect stems that are often shorter than those of *Sasa*. The arrangement of leaves is similar, as are the thickened nodes and the sheaths that remain hanging for a long period. The branches are single, with in some species two or three branches per node. *Sasaella* quickly develops many rhizomes and grows sometimes more rampantly than *Sasa*. The most widespread bamboo for ground cover is *Sasaella ramosa*. Most are hardy.

Sasamorpha

Sasamorpha Nakai, with four species from East Asia, is most suited for planting underneath trees and shrubs. Unlike *Sasa* and *Sasaella* this genus has a monopodially-branched rhizome and the culms are always well separated. In *S. borealis* the leaves develop a dry white border, similar to *Sasa*. *Sasamorpha* does best in semi-shade but will tolerate full shade.

Semiarundinaria

Semiarundinaria Nakai comes from East Asia and has about 20 species. It is fairly clump forming and has relatively large leaves. The most widespread species, *S. fastuosa*, is called 'stately bamboo' because of its straight and strong culms. It is particularly suitable for hedges. In temperate climates it grows to about 7 m (23 ft), but in its native Japan it reaches 15 m (50 ft).

The culms have a partly-developed sulcus. These are found only on those internodes that have developed branches and there only on the lower third. It too develops branch buds inside the shoot. However, in the growth period, the branches spread out so quickly that the groove forms only in part of the internode. In *Semiarundinaria* the culm sheaths remain attached only for about two weeks. In this genus each node has one main branch with a smaller branch at either side. More branches develop during the year, and if clipped short, the remaining branches grow bushier.

Shibataea

Shibataea Nakai from China and Japan has five species. It has rather unusual short, broad leaves. The leaves are attached at the node on very short stems, giving a rosette effect. In good conditions it will grow to 1.5 m (5 ft), but more usually to about 80 cm (31 in.). Their small size makes them good house plants for pots or containers. *Shibataea* dislikes strong sunlight because the leaves then lose too much water. Semi-shade and a relatively damp soil is ideal. It has long been known in botanic gardens but is rarely seen in private gardens, in the shape of the single species *S. kumasasa*. In France bamboo enthusiasts have propagated the variety *S. kumasasa* var. *aureostriata*, with yellow-striped leaves.

Sinobambusa

In European and American gardens, only one species of this tropical genus *Sinobambusa* Makino is found, namely *S. tootsik*. It is not winter-hardy but is very good as a container or house plant. In S. China, where it is native, it is often called 'temple bamboo' because it is commonly planted in temple gardens.

(*Right*) *Sasa palmata* f. *nebulosa* next to a garden pond

Sinarundinaria murielae in autumn

Sinarundinaria

The genus *Sinarundinaria* is sometimes known as *Fargesia* and is often offered under this name in catalogues (see page 124). The genus comes from central China and the Himalayas. They are usually fine-leaved bamboos with thin, bendy stems that turn yellow, green or purple, depending on age and species. They are strongly clump forming and after a few years grow to a thick bush from which the leaves hang down. Many soft, thin twigs grow from each node and these branch repeatedly over the following years. This results in the leaves and twigs getting relatively heavy and the stems bending under their weight, and in some cases even intertwining. *Sinarundinaria* is therefore very suitable as a decorative container plant. If the container is big enough *Sinarundinaria* grows quickly into a tall and attractive plant.

S. nitida is a speciality. It produces only culms in the first year, twigs and leaves not appearing until the following season. All *Sinarundinaria* species grow best in climates which are not too hot but which are relatively moist. The two species available are very hardy and will tolerate frosts of up to −20°C.

Thamnocalamus

The genus *Thamnocalamus* Munro is a good example of the uncertainty surrounding the identification and classification of particular species. Most of the species now under *Thamnocalamus* were previously included in *Arundinaria*.

They are high-altitude bamboos, usually having comparatively small, thin leaves and gently arching culms. Although they are temperate bamboos, not all species are hardy.

54

5

Species and Cultivars for the Garden

Introduction

This list of species and cultivars of horticultural bamboos can be only a snapshot. In the last couple of years there has been a big increase in the availability of bamboos. China has opened her borders and gradually new species are appearing that have not been cultivated before. These must first be verified and propagated and there is a delay of some years before they are commercially available. A few plants have also come from Central and South America, albeit rather intermittently. It remains to be seen whether these will become established here. As we have seen, the shape and growth of bamboos are strongly influenced by their environment, in particular whether they are in an oceanic or continental climate.

Details of hardiness and height in the list have been gleaned from the literature and from the research of Werner Simon, modified by our own experience. Because many species have been in cultivation for a relatively short time, we have no definitive information on growth potential, especially since they often grow more slowly under sub-optimal climatic conditions. A species that is fully hardy in the south can often freeze back and remain much smaller in regions with harder winters. The reader should appreciate that the relative lack of precise information is partly a consequence of this climatic influence on growth. The authors and publisher would welcome more information so that any new edition can be improved.

Arundinaria

In the genus *Arundinaria* Michaux there are lepto-morph species and also species in which the rhizome neck (point of attachment of the rhizome) is considerably elongated, allowing a different method of spreading. Those species which are clump forming are treated here as *Thamnocalamus* since some of these have been assigned to that genus on the basis of flower structure.

They have three to six branches per node, often not until the second or third year. Leaves in *A. gigantea* are up to 20 cm (8 in.) long and rather broad, in *A. anceps* narrow and short.

Arundinaria anceps Mitford

Origin: India, NW Himalaya to 3,500 m (11,500 ft), Sikkim, Bhutan, Gharwal
Site: semi-shade, moist, sheltered
Hardiness: to −18°C
Leaf: 10 cm (4 in.) long, narrow
Branches: at first 3–4, later many
Culm-sheath: green-beige, quick drying
Culm: shiny green to matt green-brown
Height: 1.5–4 m (5–13 ft)
Spread: spreading rapidly, making small clumps
Use: container or small shrub in shade/bushes
Speciality: thick, fresh green foliage

Arundinaria funghomii McClure

Origin: China
Site: sun to semi-shade
Hardiness: to −18°C
Leaf: 18 cm (7 in.) long, 3 cm (1 in.) wide
Branches: at first 3–4, later more
Culm-sheath: green-beige, quick drying
Culm: shiny green with orange
Height: 1.5–2 m (5–6½ ft)
Spread: separated individual stems
Use: mixed borders
Speciality: light-green foliage; interesting shape

Arundinaria gigantea subsp. **tecta** (Walter) McClure

Origin: USA
Site: sun, moist soil
Hardiness: to −23°C
Leaf: 25 cm (10 in.) long, 4 cm ($1\frac{1}{2}$ in.) wide, light green, softly hairy on both sides
Branches: acute-angled to stem, growing upward
Culm-sheath: red
Culm: green to pale yellow
Height: 2 m ($6\frac{1}{2}$ ft)
Spread: culms sometimes in tufts, spreading rapidly
Use: low-growing bank stabilizer, low hedge
Speciality: strong growing attractive foliage

Bambusa

A tropical genus with thick clumping growth and typical pachymorphic rhizome. No obvious tesselation on leaves. Half-hardy, therefore cold or temperate greenhouse necessary for over-wintering.

Bambusa glaucescens (Willdenow) Siebold ex Munro (syn. *B. multiplex* (Coureiro) Steudel

Origin: China
Site: sun
Hardiness: to −13°C
Leaf: 15 cm (6 in.) long
Branches: several per node
Culm-sheath: brown-green
Culm: matt yellow
Height: 8 m (26 ft)
Spread: clump forming
Use: container, conservatory
Speciality: stately plant with typical 'bamboo look'.
Also offered under the name 'Golden Goddess'.

Bambusa glaucescens 'Alphons Karr'

Origin: China
Site: sun
Hardiness: to −13°C
Leaf: 15 cm (6 in.) long
Branches: several per node

Culm-sheath: brown-green, white and yellow stripes
Culm: at first pink, then bright yellow to orange with green stripes
Height: 8 m (26 ft)
Spread: clump forming
Use: container, conservatory
Speciality: coloured stems

Bambusa glaucescens 'Fernleaf'

Origin: China
Site: sun
Hardiness: to −13°C
Leaf: up to 20 per branch, 15–30 mm ($\frac{3}{5}$–$1\frac{1}{5}$ in.) long
Culm-sheath: brown-green
Culm: matt yellow, not hollow, therefore also known as 'Solida'
Height: 3 m (10 ft) when planted out, in pot about 30 cm (12 in.)

Bambusa glaucescens 'Golden Goddess' as a house plant

Bambusa glaucescens 'Fernleaf' as a container plant

Spread: clump forming
Use: container, conservatory, pot-plant
Speciality: resembles fern with its 20 leaflets in two
   rows

Bambusa glaucescens var. **rivierorum**
   Maire

Origin: China
Site: sun
Hardiness: to −13°C
Leaf: up to 12 per branch, 5 cm (2 in.) long
Culm-sheath: brown-green
Culm: matt yellow
Height: 4 m (13 ft)
Spread: clump forming
Use: container, conservatory
Speciality: more elegant than the species, attractive
   foliage

Bambusa ventricosa McClure

Origin: S. China
Site: light
Hardiness: to −13°C
Leaf: 12 cm ($4\frac{1}{2}$ in.) long, 1.2 cm ($\frac{1}{2}$ in.) wide, up
   to 13 leaves per branch, up to twice as large at
   the stem tip
Branches: spreading
Culm-sheath: green, orange when dying
Culm: green cylindrical
Height: 2.5 m (8 ft) (0.8 m [$2\frac{1}{2}$ ft] when restrained)
Spread: clump forming
Use: pot or container plant, conservatory
Speciality: under certain conditions develops the
   thick internodes that give it its name 'Buddha's
   belly'

Bambusa vulgaris Schrader ex Wendland

Origin: S. China, widespread in cultivation
Site: light
Hardiness: to 0°C
Leaf: 25 cm (10 in.) long, 4 cm (1½ in.) wide
Branches: conspicuously long
Culm-sheath: wide, dark-brown hairs
Culm: green, cylindrical
Height: 15 m (49 ft)
Spread: clump forming
Use: container plant, conservatory
Speciality: thick stems, strong growth, only for large areas and containers

Bambusa vulgaris 'Vittata'

Somewhat smaller, with yellow stems, irregularly striped with green.

Chimonobambusa

The thin sheaths fall away within a year and the blades are very small. Many branches per node, sprouting in autumn and winter, branch development in early spring.

Chimonobambusa marmorea (Mitford) Makino

Origin: Japan
Site: semi-shade to sun
Hardiness: to −18°C
Leaf: 15 cm (6 in.) long, close together
Branches: short
Culm-sheath: pale green, developing white, brown marbling. The base of the sheath with an obvious ring of hairs
Culm: light green to matt red, in sun bright red
Height: 1.5 m (5 ft)
Spread: spreading rapidly in good growing conditions
Use: courtyards, conservatories, pots
Speciality: conspicuous attractive sheath

Chimonobambusa quadrangularis (Farelly: *The Book of Bamboo*)

Chimonobambusa marmorea 'Variegata'

Like the species, but with white stripes on the leaves; best sited in semi-shade.

Chimonobambusa quadrangularis (Fenzi) Makino
(syn. *Thamnocalamus quadrangularis*)

Origin: China, Taiwan
Site: sun to semi-shade
Hardiness: to −13°C
Leaf: 15 cm (6 in.) long, 1.2 cm (½ in.) wide, 5–7 leaves per branch
Culm: green to brown, rectangular in thick stems
Height: 13 m (43 ft), in containers 1–3 m (3–10 ft), depending on root space
Spread: spreading rapidly in warmer climates
Use: pot or container plant
Speciality: unusual culm shape, pretty foliage

Chusquea

This very variable genus comes from South America. Culms solid. Multiple branches, rarely just

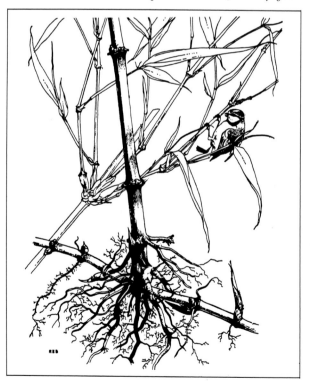

Young plant of *Chusquea coronalis*

three. Central branch sometimes longer than side branches. The latter also arise direct from the node and almost surround the node. Small narrow leaves.

Chusquea coronalis Soderstrom et Calderon

Origin: Guatemala, Costa Rica
Site: cool, moist, full shade

Hardiness: to −7°C
Leaf: very small, soft leaves
Branches: up to 50, surrounding stem at node
Culm: green
Height: 2.5 m (8 ft)
Spread: clump forming
Use: delicate pot-plant
Speciality: ring of branches surrounding nodes

Chusquea couleou E. Desvaux

Origin: Chile

Site: sun to semi-shade, moist, nourishing, fresh soil

Hardiness: to −13°C in continental climate, to −18°C in oceanic

Leaf: small blue-green leaves

Branches: many, direct from nodes

Culm-sheath: adheres long, branches appear after extension

Culm: looks like bottle-brush because of large number of branches

Height: 6 m (20 ft)

Spread: clump forming with open habit

Use: as individual plant in oceanic climate

Speciality: difficult species with rosette branches

Drepanostachyum

Clump forming genus. Culms arching in upper part. Sheath blades pointing downwards. Many thin branches, half encircling node. Leaves with obvious stalk, noticeable ligule.

Drepanostachyum falcatum (Nees von Esenbeck) P.C. Keng (syn. *Thamnocalamus falcata, Chimonobambusa falcata*)

Origin: Sikkim, Bhutan to 2,300 m (7,500 ft), India

Site: warm, semi-shade to shade

Hardiness: to −13°C

Leaf: 15 cm long

Branches: many narrow branches

Culm-sheath: dark red

Culm: grey-green, thin, upright to slightly arched

Height: 6 m (20 ft)

Spread: clump forming

Use: container plant

Speciality: extensions of sheaths with shiny inner side. Leaves often much larger at culm tip

Drepanostachyum falconeri J.D. Hooker ex Munro
(syn. *Arundinaria falconeri*)

Origin: India

Site: semi-shade

Chusquea couleou in the open, shortly after sprouting

Hardiness: to −13°C

Leaf: 10 cm (4 in.)

Branches: many

Culm-sheath: matt pink to straw-coloured, very shiny inside

Culm: olive green, matt yellow in sunny conditions, purple stain below nodes

Height: 8 m (26 ft) when sufficient room for roots

Spread: clump forming

Use: container, conservatory

Speciality: culm nodes dark purple, attractive foliage

Drepanostachyum hookerianum (Munro) P.C. Keng (syn. *Arundinaria hookeriana*)

Origin: India

Site: sun to semi-shade

Hardiness: to −18°C

Leaf: 7–30 cm (3–12 in.) long

Branches: many
Culm-sheath: pink-green, dry remnants remain long on stem
Culm: long stripes in pale yellow, pink, pale green, creamy white
Height: 6 m (20 ft)
Spread: clump forming
Use: container, conservatory
Speciality: very colourful growth, young stems deep pink in sun

Indocalamus

Sheaths remaining on culm, nodes not swollen. Auricles few. Leaf large and tough.

Indocalamus latifolius (McClure et P.C. Keng)

Origin: Nanking, China
Site: semi-shade to sun, not too hot
Hardiness: to −23°C
Leaf: large, shiny
Branches: individual, rising upwards
Culm-sheath: reddish at first
Culm: thin, green
Height: 1.5 m (5 ft)
Spread: short rhizomes
Use: good between higher decorative shrubs

Indocalamus tesselatus (Munro) P.C. Keng (syn. *Sasa tesselata* (Munro) Makino et Shibata)

Origin: China
Site: semi-shade to full sun
Hardiness: to −23°C
Leaf: very big, shiny matt, dark green
Branches: individual
Culm-sheath: red, quickly falling
Culm: thin, green, curved down by weight of leaves
Height: 1 m (3 ft)
Spread: moderate spread, loose habit
Use: ground cover under trees, high shrubs, pot-plant
Speciality: has largest leaf of all hardy cultivated bamboos

Drepanostachyum hookerianum has particularly attractive culms

Otatea

From Mexico and Honduras, sympodial (new rhizome growth at the side), three branches per node at first, later more.

Otatea aztecorum McClure et E.W. Smith

Origin: Mexico
Site: very bright, warm
Hardiness: to −3°C
Leaf: narrow, 30 cm (12 in.) long, 5 mm ($\frac{1}{5}$ in.) wide, light green
Branches: thin, lightly arched

Indocalamus tesselatus

Culm-sheath: light green with white hairs
Culm: light green, with white bloom at first, round
Height: 8 m (26 ft)
Spread: clump forming
Use: container, conservatory, also needs a lot of light in winter
Speciality: very narrow leaves, soft appearance

Phyllostachys

Sheath falling quickly, culms growing from buds on rhizome, sometimes also from the tip. All internodes of rhizome, culm and branches grooved or flattened above the bud. Each node has two branches of unequal size, sometimes with a weaker one between.

All *Phyllostachys* like the same conditions: in cold regions, at the limit of frost tolerance, they are best sited so that they get the full warmth of the sun throughout the summer, but so that they are partly shaded in winter by a building or another plant. In milder climates *Phyllostachys* needs full sun, as long as there is a good water supply.

In all species the leaves average 10 cm (4 in.) long but are very variable, especially in young plants. We have therefore simply used the comparative terms 'large' and 'small'.

The heights given are those attainable when the climate is not too cold. In continental areas with hot summers the culms harden more, and grow more upright and taller (first height given below) than in oceanic areas with mild winters and cool summers, where they remain weaker and

Otatea aztecorum

smaller, and are more arched in growth (second height given below).

Phyllostachys angusta McClure

Origin: Chekiang, China
Hardiness: to −23°C
Leaf: small, obvious ligule, rarely auricles, no bristles
Culm-sheath: cream-coloured, lavender tinted, few brown spots
Culm: upright, green, shiny, node thicker than sheath
Height: 7 m/2 m (23 ft/6½ ft)
Spread: short rhizomes
Use: grove, hedge
Speciality: elegant shape, dense wood

Phyllostachys arcana McClure

Origin: Anhwei, China
Hardiness: to −18°C
Leaf: without auricles, bristles or hairs
Culm-sheath: lavender-beige, green veins, sparsely spotted with red-brown, ligule rising above edges, blade ribbon-like, lightly wavy, bent backwards
Culm: green, powdered white at time of sheath fall, not hairy, usually noticeably ribbed, node-ring and sheath-ring projecting, old culms sometimes with black spots
Height: 9 m/4 m (30 ft/13 ft)
Spread: moderate

Phyllostachys aurea Carrière ex A. et C. Rivière

Origin: China

Hardiness: to −18°C in continental climate, to −23°C in oceanic

Culm-sheath: olive with pink, veins reddish or pale green, blades narrow and hanging down

Culm: green, going dull yellow in sun, basal internodes thickened, some nodes angled

Height: 10 m/4 m (33 ft/13 ft)

Spread: thickly clumped culms with short rhizomes in cooler climates

Use: hedge and individual plant, container

Speciality: thickening of basal internodes resembles the god of happiness, Hotei. Chinese name 'Hotei Chiku'

Phyllostachys aurea 'Holochrysa'

Culm: yellow to orange, irrespective of amount of sun. Very good container plant.

Phyllostachys aurea growing as a container plant

Phyllostachys aureosulcata – green culm with yellow sulcus

Phyllostachys aurea f. albo-variegata

Like the species, but smaller, leaf variegated with white.

Phyllostachys aureosulcata McClure

Origin: Chekiang, China

Hardiness: to −30°C

Leaf: loose-growing, small

Culm-sheath: pale olive, striped red and cream, without spots, well-developed auricle with few bristles, large ligule, blade lanceolate, hanging down

Culm: erect, matt green when young, rough, sulcus yellow, later completely yellow, sometimes strongly zigzagged near culm base

Height: 10 m/5 m (33 ft/16 ft)

Spread: long rhizomes, loose-growing culms

Use: grove, windbreak, as bushes on mixed borders

Speciality: culm habit and coloration

Phyllostachys bambusoides Siebold et Zuccarini

Origin: China
Hardiness: to −18°C in continental climate, to −23°C in oceanic
Leaf: very variable, large
Culm-sheath: green, heavily spotted with brown, auricles with large bristles, ligule large, blade upright, green with beige edges
Culm: dark, shiny green, smooth, no bloom
Height: 10 m/6 m (33 ft/20 ft)
Spread: long rhizomes when warm enough
Use: large grove

Phyllostachys bambusoides 'Castillonis'

Like species, but culm bright yellow with green sulcus, some leaves with single white stripes. New growth very colourful, with pink to orange-red leaf sheaths. Shape very attractive. Valuable cultivar, almost disappeared after 1966–8 flowering.

Phyllostachys bambusoides 'Katashibo'

Like species, but internodal groove is 'wrinkled' with longitudinal ridges.

Phyllostachys bambusoides 'Violascens'

Like species, but with shorter internodes. Culm finely striped, at first green and yellow with orange, later turning to bright purple-violet, especially in shade. Old culms also have short, white marks and, in sun, dense-brown spots.

Phyllostachys bambusoides 'White Crookstem'

Like forma *geniculata* but with white mealy powder, almost covering the green culm in older plants.

Phyllostachys bambusoides. Species and cultivars of this genus are good for planting as a grove

Phyllostachys bambusoides f. geniculata
(Nakai) Muroi (syn. *P. b.* 'Slender Crookstem')

Like species but culm base sometimes curved, then erect again. Nodes not as raised.

Phyllostachys bissetii McClure

Origin: Sichuan, China
Hardiness: to Phyllostachys bambusoides30°C
Culm-sheath: light to yellow-green, red tint, thin and easily split
Culm: gently arched
Height: 7 m/3 m (23 ft/10 ft)
Spread: moderately spreading
Use: grove, hedge
Speciality: very hardy and decorative

Phyllostachys decora McClure

Origin: Kiangsu, China
Hardiness: to Phyllostachys bambusoides23°C
Culm-sheath: dark green, unmarked or with fine dark spots, pale stripes, red margin to tip, blade red, erect
Culm: erect
Height: 6 m/4 m (20 ft/13 ft)
Spread: moderately spreading
Use: individual plant, grove
Speciality: pretty, colourful young growth

Phyllostachys flexuosa (Carrière) A. et C. Rivière

Origin: China
Hardiness: to −23°C
Culm-sheath: green-beige with close red veins and small brown speckles. Ligule large and dark, blade narrow and bent back
Culm: green, later yellow with black speckles getting larger with age, zigzag growth, upright or slightly arched
Height: 10 m/2 m (33 ft/6½ ft)
Spread: moderately spreading

(*Left*) *Phyllostachys bambusoides* 'Castillonis' – yellow culm with green sulcus

Use: grove, bush and container plant
Speciality: elegant culm and branches

Phyllostachys hetrocycla f. pubescens
(Houzeau de Lehaie) Muroi in Sugimoto (syn. *P. pubescens*)

Origin: China
Site: warm
Hardiness: to −23°C
Leaf: small to quite large
Culm-sheath: green-grey, densely speckled with black, brown hairs, auricles with long bristles and a large ligule with long dark-red fringe
Culm: grey-green, with soft white hairs when young
Height: 8 m/3 m (26 ft/10 ft)
Spread: spreading when kept in warm conditions
Use: container plant, solitary
Speciality: important food plant in China, does not grow large in temperate regions

Phyllostachys humilis Muroi

Origin: China
Hardiness: to −23°C
Culm-sheath: light green with red veins, thick covering of white hairs, large, dark ligule
Culm: light olive green, branches almost horizontal
Height: 5 m/3 m (16 ft/10 ft)
Spread: moderately spreading
Use: grove, container, bonsai
Speciality: used for bonsai in Japan

Phyllostachys meyeri McClure

Origin: Chekiang, China
Hardiness: to −18°C
Culm-sheath: matt green, speckled with brown, ring of white hairs at base, otherwise hairless, no auricles or bristles, blade wavy
Culm: erect
Height: 11 m/5 m (36 ft/16 ft)
Spread: moderately spreading
Use: grove

Phyllostachys nidularia Munro

Origin: China
Hardiness: to −18°C
Leaf: large, light green
Culm-sheath: green, large auricles reaching the blade
Culm: sheath ring hairy at first, node ring thick
Height: 8 m/3 m (26 ft/10 ft)
Spread: moderately spreading
Use: good, dense decorative container plant
Speciality: very early, tasty shoots, best for cooking

Phyllostachys nidularia 'Smoothsheath'

Like species, but with bare sheath ring and hairless culm sheath.

Phyllostachys nigra (Loddiges ex Lindley) Munro

Origin: China
Site: warm, sunny
Hardiness: to −18°C in continental areas, −23°C in oceanic
Leaf: small
Culm-sheath: pinky-beige with dark spots in the upper part, blade small, green and pleated, ligule with fringe and large auricles with red to dark-red bristles
Culm: green at first, gradually becoming shiny black, arched
Height: 5 m/2.5 m (16 ft/8 ft)
Spread: very moderate spread
Use: individual plant, container plant
Speciality: shiny black culms

Phyllostachys nigra 'Fulva'

Like *P. nigra* but taller, culm brown with dark spots, edge of leaf dark brown.

Phyllostachys nigra f. **boryana** (Mitford) Makino

Origin: China
Hardiness: to −23°C
Culm-sheath: like *P. nigra*

Culm: not black but speckled with brown, erect
Height: 15 m/4 m (50 ft/13 ft)
Spread: moderately spreading
Use: individual plant
Speciality: decorative culm colouring

Phyllostachys nigra f. **henonis** (Mitford) Muroi

Origin: Guangdong, Sichuan, China
Hardiness: to −23°C
Leaf: shiny light green, somewhat wavy
Culm-sheath: tan coloured, tinted pink, blade short, wavy, dagger-shaped
Culm: green to yellow, mealy in hot summer weather, gently arched to erect, right to the tip
Height: 16 m/5 m (52 ft/16 ft)
Spread: moderately spreading
Use: solitary plant, with shrubs
Speciality: shape strongly influenced by climate

Phyllostachys nigra (Farelly: *The Book of Bamboo*)

Phyllostachys nigra f. **megurochiku** Makino ex Tsuboi

Like *P. nigra* f. *henonis*, culm green, sulcus becoming dark brown.

Phyllostachys propinqua McClure

Origin: Kwangsi, China
Hardiness: to −30°C
Leaf: dark green
Culm-sheath: pale olive with bronze, blade narrow, without auricles or bristles
Culm: bright dark green, prettily branched
Height: 12 m/6 m (39 ft/20 ft)
Spread: moderately spreading
Use: individual plant, container plant

Phyllostachys purpurata McClure

Origin: Kwangtung, China
Hardiness: to −18°C
Leaf: small and light green, turning irregular yellow before falling, leaf sheath without auricles or bristles
Culm-sheath: green and bare, often bluish at base, auricles and bristles small or absent, blade red
Culm: thin, arched and markedly zigzag, especially where branched
Height: 5 m/2 m (16 ft/6½ ft)
Spread: moderately spreading
Use: grove, container plant
Speciality: colouring of leaves, culm shape

Phyllostachys purpurata f. **solida** S.L. Chen (syn. *P. purpurata* 'Solidstem')

Like the species, about 30 per cent smaller, but culm solid in lower half. Native to Anhwei.

Phyllostachys rubromarginata McClure

Origin: China
Hardiness: to −23°C
Leaf: large leaves, small auricles with short bristles, ligule with long dark-red fringe
Culm-sheath: olive green, reddish towards the tip, with dark-red border, without auricles or bristles, ligule fringed dark red, blade straight, lower ones adhering, upper ones sticking out
Culm: green to yellow, bare, sulcus hardly noticeable on unbranched nodes
Height: 8 m/4 m (26 ft/13 ft)
Spread: moderately spreading
Use: groves, hedge, container plant

Phyllostachys viridiglaucescens (Carrière) A. et C. Rivière

Origin: China
Hardiness: to −23°C or −30°C
Leaf: shiny green, undersides hairy and bluish
Culm-sheath: matt green, red-veined, speckled dark, large auricles with long bristles, long blade somewhat pleated
Culm: nodes markedly raised, stem with bluish bloom after sheaths have fallen
Height: 10 m/6 m (33 ft/20 ft)
Spread: moderately spreading
Use: bushes, groves
Speciality: widely arching stems

Phyllostachys viridis (R.A. Young) McClure

Origin: China
Hardiness: to −18°C or −23°C
Leaf: small, auricles and bristles only developed on young stems
Culm-sheath: light matt pink, brown spots, green veins, without auricles and bristles, ligule with stiff fringe, blade somewhat wavy
Culm: matt green, minutely dimpled, unbranched nodes not raised
Height: 16 m/7 m (52½ ft/23 ft)
Spread: spreading in warm climates
Use: groves
Speciality: warmth loving, forms strong culms quickly – 'thick bamboo'

Phyllostachys viridis 'Houzeau'

Like the species, but with yellow sulcus.

Arching culm of *Phyllostachys viridiglaucescens*

Phyllostachys viridis 'Robert Young'

Like the species, but smaller, culm bright golden-yellow with green stripes running down the internodes from a green ring below the nodes.

Phyllostachys vivax McClure

Origin: Chekiang, China
Hardiness: to −23°C
Leaf: large
Culm-sheath: matt cream-coloured, speckled with brown, without auricles or bristles, blade ribbon-like, heavily pleated
Culm: not completely straight, grooved, bluish at first
Height: 15 m/6 m (49 ft/20 ft)
Spread: spreading in good growing conditions
Use: groves
Speciality: elegant foliage

Pleioblastus

Culm-sheaths leathery and long-lasting, with well developed blades, several branches produced during the year at each node, puts out growth in early spring.

The genus is divided into three sections and, with the exception of *P. simonii* (section Medakea) all species listed here belong to section Nezasa. All are more or less invasive and prefer fertile, moist soils. Species with large foliage should not be placed in full sun in regions with hot, dry summers.

Pleioblastus argenteostriatus (Regel) Nakai

Origin: known only in cultivation
Site: semi-shade to full shade
Hardiness: to −23°C

Leaf: 15 cm (6 in.) long, 1.8 cm ($\frac{4}{5}$ in.) wide, irregular white stripes
Culm: green, somewhat arched
Height: 0.60 m (2 ft)
Spread: moderately spreading
Use: as individual plant for shady spots, ground cover underneath tall hedges
Speciality: dark blue-green leaves with white stripes

Pleioblastus chino (Franchet et Savatier) Makino

Origin: S. Hokkaido, N. and C. Honshu, Japan
Site: sun to semi-shade
Hardiness: to −23°C
Leaf: 20 cm (8 in.) long, 1.8 cm ($\frac{4}{5}$ in.) wide
Culm: green to brown, sheath persistent
Height: 1.5–2.5 m (5–8 ft)

Spread: rampant
Use: hedge

Pleioblastus chino f. **aureo-striatus** (Makino) Muroi et Kasahara

Origin: known only in cultivation
Site: semi-shade to full shade
Hardiness: to −23°C
Leaf: 11 cm (4 in.) long, 1.1 cm ($\frac{2}{5}$ in.) wide, fine white stripes, somewhat faded
Culm: green
Height: 40 cm (15 in.), to about 2 m (6$\frac{1}{2}$ ft) in older plants
Spread: moderately spreading
Use: as border in front of evergreens
Speciality: very delicate appearance

Pleioblastus chino f. *aureo-striatus* has fine creamy-yellow stripes on its leaves

Boldly-striped *Pleioblastus fortunei* is ideal for underplanting

Pleioblastus chino var. **gracilis** (Makino) Nakai

Origin: Honshu, Japan
Site: semi-shade
Hardiness: to −23°C
Leaf: 13 cm (5 in.) long, 1.2 cm ($\frac{1}{2}$ in.) wide, glabrous
Height: 0.6 m (2 ft), taller with age
Spread: moderately spreading
Use: as 'grass' in herbaceous beds

Pleioblastus pumilus (Mitford) S. Suzuki (syn. *Sasa pumila*)

Origin: S. Japan
Hardiness: to −23°C
Leaf: 20 cm (8 in.) long, 2.5 cm (1 in.) wide
Culm: green, reddening in the sun, obvious ring of hairs at the nodes

Height: 1.2 m (4 ft)
Spread: extremely rampant
Use: ground cover, can be mown

Pleioblastus fortunei (Van Houtte) Nakai

Origin: known only in cultivation
Site: semi-shade to full shade
Hardiness: to −23°C
Leaf: 15 cm (6 in.) long, 1.4 cm ($\frac{1}{2}$ in.) wide, hairy on both sides
Culm: green, glabrous
Height: 40 cm (15 in.)
Spread: quite vigorous
Use: ground cover, underplanting, pot-plant
Speciality: pretty, variegated, carpet bamboo

Pleioblastus fortunei is a very robust ground cover species

Pleioblastus pygmaeus var. **pygmaeus**
(Miquel) Nakai

Origin: Japan
Site: sun to semi-shade
Hardiness: to −30°C
Leaf: small blue-green leaves, some turning dry and white in autumn
Culm: sheath-ring with dense, short hairs
Height: 30 cm (1 ft)
Spread: rampant
Use: ground cover, plant with woody species and large perennials, lawn substitute, pot-plant
Speciality: marked seasonal change in appearance, pale green in late autumn

Pleioblastus pygmaeus var. **distichus**

Origin: only in cultivation
Site: sun to semi-shade
Hardiness: to −23°C
Leaf: 7 cm (3 in) long, 0.8 cm ($\frac{1}{5}$ in.) wide, rather tough and stiff, both sides glabrous
Culm: green, glabrous, stem-sheath glabrous
Height: 0.6 m (2 ft)
Spread: rampant
Use: ground cover, pot-plant

Pleioblastus simonii (Carrière) Nakai

Origin: S. Japan
Site: sun to semi-shade
Hardiness: to −23°C
Leaf: 25 cm (10 in.) long, 2.5 cm (1 in.) wide
Culm: upright
Height: 4 m (13 ft)
Spread: invasive in good growing conditions
Use: upright hedge or screen

Pleioblastus simonii 'Heterophyllus'

Leaf: variable width, some striped with white
Height: 4 m (13 ft)
Spread: rampant
Use: hedge
Speciality: different leaves on the same plant

Pleioblastus viridistriatus (Siebold) Makino (syn. *Arundinaria auricoma*)

Origin: known only in cultivation
Site: semi-shade to full sun
Hardiness: to −23°C
Leaf: 20 cm (8 in.) long, 2.5 cm (1 in.) wide, softly hairy on both sides, in spring yellow with a green stripe, getting greener later in the year
Height: 1.5 m (5 ft)
Spread: very little spread

Use: with evergreens, beneath tall trees, containers
Speciality: very pretty yellow-green foliage

Pseudosasa

Persistent culm-sheaths, branches single or up to three, on swollen nodes, rhizome sympodially branched, fresh annual growth from side buds (compare *Sasamorpha*).

Pseudosasa japonica (Siebold et Zuccarini ex Steudel) Makino ex Nakai

Origin: S. Japan, Korea
Site: sun or shade
Hardiness: to Pseudosasa23°C
Leaf: up to 30 cm (12 in.) long, 3 cm (1⅕ in.) wide, shiny, leathery

Pleioblastus pygmaeus

74

Pseudosasa japonica is at home in the shade

Branches: only at tip of culm
Culm: green, sheath remains attached
Height: 3.5 m (11½ ft)
Spread: moderately spreading
Use: individual, hedge, container
Speciality: P. j. tsutsumiana has bottle-shaped thickened internodes

Sasa

Persistent culm-sheaths, single branches, leptomorph rhizome, nodes markedly swollen, culmsheath usually shorter than internode.

Sasa kagamiana Makino et Uchida

Origin: N. Honshu, Japan
Site: semi-shade to full shade
Hardiness: to −23°C

Leaf: up to 30 cm (12 in.) long and 6 cm (2½ in.) wide, leathery
Culm: stem sheath with obvious white hairs
Height: 2 m (6½ ft)
Spread: spreading
Use: in shade
Speciality: young culm has noticeable white hairs

Sasa kurilensis (Ruprecht) Makino et Shibata

Origin: Korea, Japan, Sakhalin, Kuril Islands
Site: sun to semi-shade
Hardiness: to −30°C
Leaf: 20 cm (8 in.) long, 4 cm (1½ in.) wide, shiny
Branches: ending at same height as stem
Culm: glabrous, with white, mealy bloom
Height: 2.5 m (8 ft)
Spread: spreading rapidly
Use: extensive planting

75

Twenty-year-old stand of *Pseudosasa japonica*

Sasa palmata (Bean) Camus

Origin: Japan
Site: sun to full shade
Hardiness: to −30°C
Leaf: 30 cm (12 in.) long and 10 cm (4 in.) wide, leathery, glabrous
Culm: green, white mealy bloom; *S. p.* f. *nebulosa* speckled brown
Height: 2 m (6½ ft)
Spread: invasive
Use: individual, hedge

Sasa tsuboiana Makino

Origin: Honshu and Shikoku, Japan
Site: sun to full shade
Hardiness: to −23°C
Leaf: 25 cm (10 in.) long, 5 cm (2 in.) wide
Height: 1.5 m (5 ft)
Spread: moderately spreading
Use: small bush

Sasa veitchii (Carrière) Rehder

Origin: S.W. Honshu, Japan
Site: full shade
Hardiness: to −23°C
Leaf: 25 cm (10 in.) long, 5 cm (2 in.) wide, leathery, dark green, with white drying edge towards autumn
Height: 1.5 m (5 ft)
Spread: spreading
Use: underplanting, best in front of evergreens
Speciality: attractive white edging to leaves

Sasa veitchii 'Minor'

Origin: Hokkaido to Kyushu, Japan
Site: sun to full shade
Hardiness: to −23°C
Leaf: 8 cm (3 in.) long, 1.5 cm ($\frac{3}{5}$ in.) wide, drying at edge in autumn, softly hairy beneath
Height: 40 cm (16 in.)
Spread: rampant
Use: good for ground cover
Speciality: horizontal soft foliage with decorative white drying edge

Sasaella

Persistent culm-sheath and leptomorph rhizome, culms erect, nodes markedly swollen, branches mostly single, sometimes two to three.

Sasaella masamuneana (Makino) Hatusima et Muroi (syn. *Arundinaria purpurea*)

Origin: Honshu and Kyushu, Japan
Site: sun to semi-shade
Hardiness: to −23°C
Leaf: 18 cm (7 in.) long, 5 cm (2 in.) wide
Branches: single, alternate
Culm: green to red
Height: 2 m (6$\frac{1}{2}$ ft)
Spread: weakly suckering
Use: bush
Speciality: attractive red culm

Sasaella ramosa (Makino) Makino (syn. *Arundinaria vagans*)

Origin: Japan
Hardiness: to −30°C
Leaf: 15 cm (6 in.) long, 2 cm ($\frac{1}{5}$ in.) wide, white hairs, thin on upper surface, thick below
Branches: single
Culm: green, relatively thin
Height: 60 cm (24 in.)
Spread: rampant
Use: ground cover for large areas

Sasa palmata f. *nebulosa* has particularly broad leaves

Sasamorpha

Persistent culm-sheaths, rhizomes monopodially branched (fresh annual growth from rhizome tip), culms always individual.

Sasamorpha borealis (Hackel) Nakai

Origin: E. and S. Japan, Korea
Site: semi-shade to full shade
Hardiness: to −23°C
Leaf: 16 cm (6 in.) long, 2.2 cm (1 in.) wide, shiny dark green, drying white at edge in autumn
Branches: concentrated towards culm tip
Culm: sheath red, culm green to red
Height: 2 m (6½ ft)
Spread: spreading
Use: as shade plant beneath large trees

Semiarundinaria

Culm-sheaths remain attached at node for only a short time, before dropping off; internodes cylindrical, slightly flattened above the branches, three branches per node (compare *Phyllostachys*).

Semiarundinaria fastuosa (Mitford) Makino ex Nakai

Origin: S.W. Honshu, Japan
Site: warm
Hardiness: to −23°C
Leaf: 20 cm (8 in.) long, 2.5 cm (1 in.) wide
Branches: 3–8 short branches per node
Culm: erect, green to red-brown
Height: 7 m (23 ft)
Spread: spreading only in warm growing conditions
Use: columnar, tall hedge
Speciality: unusually upright for height, thick culm

Semiarundinaria yashadake (Makino) Makino

Origin: Japan
Hardiness: to −23°C

Leaf: 20 cm (8 in.) long, 4 cm (1½ in.) wide, thin, somewhat pleated, long bristles
Branches: 3–8 per node
Culm: green, sheath base hairy
Height: 4 m (13 ft)
Spread: little spread
Use: bush, hedge

Shibataea

Shibataea kumasasa (Zollinger ex Steudel) Makino ex Nakai

Origin: Japan
Site: warm, damp
Hardiness: to −23°C
Leaf: 10 cm (4 in.) long, 2.5 cm (1 in.) wide
Branches: short branches give impression of whorled leaves
Culm: green
Height: 0.8 m (2½ ft)
Spread: very little spread
Use: ground cover, pot-plant, with shade shrubs
Speciality: wide, short foliage

Sinarundinaria

Sinarundinaria murielae (Gamble) Nakai (syn. *Fargesia murielae*)

Origin: Himalayas
Site: sun to semi-shade
Hardiness: to −30°C
Leaf: 10 cm (4 in.) long, 1.2 cm (½ in.) wide
Branches: thin, 10 or more per node
Culm: arched, with white bloom in early growth
Height: 4 m (13 ft)
Spread: clump forming
Use: individual or hedge plant

Sinarundinaria murielae 'Weihenstephan'

Like the species, but stronger growth.

Semiarundinaria fastuosa (column bamboo) is very suitable for a hedge

Sinarundinaria nitida (Mitford) Nakai (syn. *Fargesia nitida*)

Origin: Himalayas
Site: semi-shade, especially in hot, dry summers
Hardiness: to −30°C, even root-balls withstand frost to −15°C
Leaf: 6 cm ($2\frac{1}{5}$ in.) long, 0.8 cm ($\frac{1}{5}$ in.) wide, dark green
Branches: many per node, thin
Culm: white bloom, bluish, later dark green and red-brown, upright
Height: 4 m (13 ft)
Spread: clump forming
Use: container, hedge
Speciality: very hardy

Sinarundinaria nitida 'Eisenach'

Site: semi-shade, especially in hot, dry summers
Hardiness: to −30°C
Leaf: 6 cm ($2\frac{1}{5}$ in.) long, 1 cm ($\frac{2}{5}$ in.) wide, dark green
Branches: many per node, thin
Culm: white bloom, bluish to dark green and red-brown, arching
Height: 4 m (13 ft)
Spread: clump forming
Use: container, individual planting
Speciality: very hardy

Sinarundinaria nitida 'Nymphenburg'

Site: semi-shade, especially in hot, dry summers
Hardiness: to −30°C
Leaf: 6 cm (2$\frac{1}{5}$ in.) long, 0.5 cm ($\frac{1}{5}$ in.) wide, dark green
Branches: many per node, thin
Culm: white bloom, bluish to dark green and red-brown, weakly arched
Height: 4 m (13 ft)
Spread: clump forming
Use: container, individual planting
Speciality: very hardy, small narrow leaves

Thalmocalamus

Thalmocalamus spathiflorus (Trinius) Munro

Origin: India
Site: bright semi-shade
Hardiness: to −13°C
Leaf: 15 cm (6 in.) thin
Branches: 2–3, later more
Culm-sheath: light green, red spots, hairy, falling
Culm: bright green, red in sunny conditions
Height: 3.5 m (11$\frac{1}{2}$ ft)
Spread: clump forming
Use: container, conservatory or outside in mild climates

Thalmocalamus tesselatus (Nees van Esenbeck) Soderstrom et Ellis (syn. *Arundinaria tesselata*)

Origin: S. Africa
Site: sun
Hardiness: to −23°C
Leaf: 12 cm (4$\frac{1}{2}$ in.), blue-green

Thamnocalamus tesselatus

Branches: many, first appearing in second year
Culm-sheath: pale green at first, later drying white, persistent
Culm: green to brown, red in sunny conditions, erect
Height: 2–4 m (6$\frac{1}{2}$–13 ft) in warm, damp sites
Spread: loosely growing culms, clump forming
Use: small shrubs, ideal near house
Speciality: very colourful, with blue-green foliage, reddish culms and white culm-sheaths

6

Planting and Cultivation

Buying a bamboo plant

Bamboo plants are offered for sale in containers in nurseries and garden centres and one can therefore, in theory, plant them throughout the year. However, a bamboo has a better chance of developing roots and surviving its first winter without damage if planted in the first half of the year. At this time the plant has so much reserve in its rhizomes that new culm growth is possible even before new roots are established. After the above-ground growth period in spring and summer, the rhizomes begin to grow and to develop roots in June and July. If the plant is already planted in its desired position at this stage it can become well established in the ground before winter. This new root growth guarantees adequate moisture supply during the following winter. The plants will be better able to withstand cold and drying winds. Bamboos planted out in autumn look more decorative, since they have last spring's culms. Most bamboos have a resting period from late summer to spring, however, and will therefore not usually produce fresh rhizomes in the autumn. Bamboos planted in the autumn therefore need close attention during the winter and must be watered to prevent drying and, in cold climates, protected from frosts. They should not be compared with other woody plants that can be planted in autumn to root vigorously in the spring. In most species, as explained on page 28, the culm develops first, followed by growth of the rhizomes.

Buying a bamboo is risky for two reasons: firstly, it is difficult to see the rhizome to check whether it is really intact and has healthy buds – a requirement for healthy growth in all bamboos. Secondly, a container plant is not always what it appears to be. For example, a few years ago beautiful Mediterranean-grown *Phyllostachys aurea*

was on offer. It later became clear that these had been taken from old, naturalised stands and 80 per cent made no further growth. It is therefore best to buy bamboos from a reputable grower, if possible from one specializing in bamboos and propagating their own 'young' plants.

It is important to buy a healthy plant that has already developed young rhizomes. It is only in the clump-forming genera, such as *Sinarundinaria*, that the aerial parts of the plant can tell you about its overall quality and vitality. Sometimes a container plant will have only one or two culms, but many rhizomes, and such a plant will develop very well in the garden. One often sees a container plant offered with both older, thick stems, several years old, and thinner, year-old stems. Such a plant will have been taken by dividing up a large bamboo plant. The thicker stems are those that grew out of the whole rhizome, whereas the thinner ones have grown up from the new rhizomes put out after division of the original plant. As long as there are plenty of fresh culms, such a plant should be perfectly satisfactory. One can tell the age from the number of branches as well as from the size of the culms. The older the plant, the more branching there is. As a rule, try to buy bamboo plants that are reasonably young, with well developed rhizomes and young culms. Large, old bamboos sometimes do not develop any better than young ones after planting.

Siting

The position for a bamboo should be chosen primarily for its visual effect. It should be decorative and harmonize with the other plants. However, it is important not to forget the plant's requirements. Bamboos are very sensitive to climate and water regime. Many species like full sun, whilst others prefer shade or semi-shade. Ground cover species such as *Sasa*, *Sasaella* and *Pleioblastus* can even tolerate deep shade. If no attention is paid to these requirements then disappointment is inevitable. An attractive *Phyllostachys* on a terrace in a shady, cool corner might look particularly decorative but will soon lose its looks and fail to put out fresh culms because it is not warm enough. *Sasa* species are ideal as ground cover but can become leggy and unattractive if planted in full sunlight. In their native countries and in oceanic climates these species can tolerate hotter sun because the air humidity is higher. In a dry warm climate, drought damage can set in, the plants becoming yellow and eventually dying off.

No bamboo likes a strong wind and the site must be selected with this fact in mind. Bamboos are thirsty plants and dry out quickly in the wind. They do excellently when grown in protected positions such as against a wall, in a corner or in a courtyard, where a wall stores the heat of the day and continues to release it in the evening, effectively producing a warmer climate. Bamboos often grow with tropical luxuriance in sheltered yards in cities. Here they are protected from wind and surrounded by walls that give out warmth, and they are not subjected to the extremes of winter cold. Bamboos, like all plants, are strongly influenced by their microclimate. They do well in courtyards or in a garden with a surrounding hedge. Several species are suitable as hedge plants but you do not need to worry about the wind if the hedge itself is bamboo. Their thick foliage protects them from drying out.

Even those species that like full sun in summer need to be provided with some shade in winter. Winter sun combined with frost is bad for bamboos, as for all evergreens. Whenever possible plant bamboos so that the low winter sun creates at least partial shade, for instance by a tree. Even if the tree is leafless, its branches and twigs will

A hemp rope between two posts supports tall culms

provide sufficient shade to prevent damage to the bamboo. When planting bamboos do not forget that they can grow quite large, in width as well as height. An individual plant such as *Sinarundinaria* can easily grow to a diameter of more than two metres (six feet) in a few years and this should be borne in mind when selecting the site.

For invasive species one must also consider the underground growth potential. A strongly rampant bamboo in the middle of a lawn is not exactly what every gardener dreams of, nor does one usually want bamboos springing up in every flower bed. If you have a garden pond you need to ensure that the rhizome tips do not bore through the lining. This danger can be diminished by using a lining material thicker than 0.5 mm ($\frac{1}{5}$ in.). Lastly, do not forget that your neighbour might not want bamboos spreading into his garden!

Soil

Bamboos are not particularly fussy about soil type. In very heavy soils or in pure sand they do not grow as well, but soils can always be improved. However they do dislike soggy soil. Take care, therefore, that you do not plant bamboos near to the water table, nor in soils where the water flow is restricted. One often finds, particularly in new housing developments, that the gardens consist of just a thin layer of soil on top of a layer of heavy soil that has been compacted by building machinery. Rainwater collects in such soil and bamboos dislike these conditions. The rhizomes rot and the plants die. A humus-rich layer of soil 30–50 cm (12–20 in.) deep over loamy soil is ideal but not always available. One therefore often has to do some soil preparation to create optimal conditions for bamboos.

If the soil is very porous, bamboos must be watered regularly because they take up a lot of water. Like all shallow-rooted species and all grasses bamboo has a rapid water metabolism.

Where the soil is very compacted, for example by building machinery, it has to be loosened up, usually by deep digging to break up the dense

Planting a divided *Phyllostachys aurea*

layers. This goes not just for bamboos, but for all other large plants.

If you want to plant bamboo alongside water, where it looks marvellous, make sure that the bank is at least 30 cm (12 in.) higher than the normal water level of the stream or pond.

How to plant

When you have purchased container plants from a nursery and chosen the site with care it is time to move on to the actual planting, which is simplicity itself. Dig a hole slightly wider than the container and see whether further digging is needed to improve drainage. Now remove the plastic container prior to planting. In very light

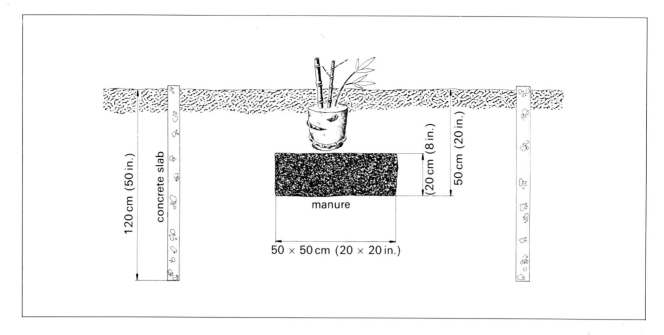

120 cm (50 in.)

concrete slab

20 cm (8 in.)

50 cm (20 in.)

manure

50 × 50 cm (20 × 20 in.)

When transplanting a bamboo from a container to the garden make sure it is planted no deeper than it grew when in the pot, unless you are putting it in very light soil, in which case it can be planted a little deeper. Put a layer of manure beneath the plant and mulch the surface to reduce evaporation. Invasive species can be held in check by concrete slabs or by a wall

soil plant the bamboo slightly deeper than it was in the container and cover the surface with light garden soil. This delays for a few weeks the rhizomes pushing up to the light and forming culms. On the other hand, if your soil is relatively heavy, you should plant the bamboo about as deep as it was in the container. Bamboos grow larger and more powerfully in their native sites, and there new rhizomes can push up through thick layers of soil. However, one cannot depend upon this kind of growth with cultivated garden bamboos.

After planting, water well and firm down the soil. When planting in spring, however, be careful not to tread on any fresh rhizomes that may have sprouted a few centimetres above the surface. Whatever the season, the soil should be mulched to reduce water loss and promote growth. Well-rotted compost is suitable, or fallen leaves from trees and bushes (but not sweet chestnut leaves or those from nut trees, since these contain a lot

of tannic acid). Peat is not suitable because it tends to draw the water out of the soil and so has the opposite effect. If you use freshly cut grass, mix it with chopped-up twigs otherwise it will get too hot as it rots. Sawdust, hay and straw are also useful. Let the mulch material settle into the plant from above to a depth of 20 cm (8 in.). The procedure is similar for planting in a wooden tub or stoneware container, for example. Make sure that the container is large enough to allow for sufficient root development for the plant height desired. If it is too small you may have to re-pot the plant the following year, otherwise it will outgrow the pot and begin to starve because it will have developed too many culms. It is particularly important that the container has sufficiently large drainage holes.

Replanting bamboos

Occasionally you might wish to move a well-established bamboo – perhaps because the site is not after all appropriate, or because you have rearranged the garden, or, as occasionally happens, you are moving house and want to take it with you. However remember that moving a bamboo plant weakens it. The rhizomes and fine roots are always damaged, even if the utmost care is taken. Therefore one needs to choose the best time for the transfer, and the best time is either shortly before the new growth begins, or immediately after the new culms have completed their growth. This is the moment before the rhizome starts its own growth. Rhizome growth should not be interrupted by replanting or division. It is generally better not to move the whole plant to the new site but to take pieces with just a few culms, or, for example with *Pleioblastus*, to plant pieces broken off the rhizome. This gives rise to more vigorous plants than replanting the whole bamboo. Sometimes no more new culms develop in the centre because the older rhizomes have been damaged during the transfer, and the new rhizomes develop outwards.

After transplanting it helps to keep the plant in still air and shade under a sheet for about a week, and to reduce the leaf area by cutting back the side branches by one to two thirds. This is important because the developing rhizome cannot take up a lot of water from the soil.

Well-developed plants often take several years to return to their original size and beauty after transplanting. However this is not always the case, and much depends upon the condition of the plant, on the species and on the precise environment of the new site.

Watering and feeding

Bamboos are very thirsty and hungry plants. They need a lot of water because their many very thin leaves lose a lot of moisture by transpiration. Garden bamboos cannot rely entirely on rain for their moisture. In their native habitat it rains more often and throughout the year and the air is humid. Bamboos therefore need to be watered in periods of extended drought and between frosts in the winter. If a bamboo gets too dry the leaves roll up as a protection against drought. By rolling up, the leaves reduce their surface area and lose less water by transpiration. If you water a bamboo plant that has rolled up its leaves it will unroll them and reveal its shiny leaves once more. Bamboos do not dry out immediately like some

of our other garden plants. Some species from the cloud zone of the Himalayas (e.g. *Sinarundinaria*, *Arundinaria* and some *Thamnocalamus*) roll up their leaves in full sun even if there is sufficient ground water, during hot summer weather; this is their means of protecting themselves.

But it is not enough simply to watch the leaves. You should try to judge how dry the soil is in the garden or container and whether and when it is necessary to water. In the case of containers especially, you need to develop sensitive fingertips to check soil moisture. It is very easy to overdo the watering and if waterlogging sets in then the rhizomes and roots rot and the plant may die.

If you are the proud owner of a bamboo hedge you can make the watering easier. Simply lay a spray pipe between the plants, or use a hose that you have punctured at regular intervals with small holes. The whole hedge can be watered thus without one having to stand and spray. Larger groups of bamboos are not so likely to suffer, because they create their own shade.

Bamboos need a lot of fertilizing. In Japan valued bamboo plantations are fertilized with horse manure, which contains a very high proportion of nitrogen. If you have access to half-rotted horse or cow manure you should apply liberal quantities in autumn or early in the spring. Riding schools are often glad to give away manure since they have no use for it. The chance should be grabbed. If you compost fresh manure in summer it is ready for use the following spring. If it is too concentrated it will not rot, so it is a good idea to mix it with straw. Then it rots quickly and completely. If there is no animal dung available you can use fertilizers with a high nitrogen content. These also have the necessary phosphorus and potassium (NPK-fertilizers).

An alternative to these commercial fertilizers are plant infusions of the herb comfrey. It is simple to make such an infusion: half fill a barrel or bucket with freshly cut plants and fill it up with water. Leave the mixture in the sun for about ten days until it begins to ferment. Stir repeatedly to get oxygen into the mixture and speed up the decomposition (the nasty smell can be reduced by extract of valerian). The fermented plant liquid manure is diluted with eight times the quantity of water and poured onto the bamboo plants. If new shoots have sprouted, dilute the manure even more otherwise it might burn the tender new growth. This method is particularly good for ground-covering bamboo species that are not easily covered by farmyard manure.

Fertilizing with nitrogen makes the plants grow strongly. Apply in summer when the rhizomes develop and again in autumn and you will have good strong plants. However, rich applications of fertilizer make the cells bigger and the cell walls thinner. Such plants lose some of their hardiness, at least as far as the stems and leaves are concerned. Therefore nitrogen-rich fertilizers should only be applied to outdoor bamboos until July or August. The plants can then ripen peacefully in late summer, the stems can grow, and the leaves are better protected against the cold. There is no difficulty with those grown indoors, or those in containers that are brought inside during the winter. However, even these will probably benefit from a period without fertilizers – like all plants.

Silica (SiO_2) is just as important for bamboos as nitrogen. Bamboo plants contain a large amount of silica and this must be supplied continuously. When growing in large groves bamboos get sufficient silica naturally by recycling it from their shed leaves, which rot on the soil below and release it. One should therefore let fallen leaves accumulate beneath individual bamboo plants. Rake together those that have fallen on the soil around the plants and place these under the plants as well. This should give the bamboos enough silica, but it is a good idea to supplement this in the first few years with SiO_2. This can be supplied in organic form as Horsetail extract. Horsetail contains even more silica than bamboo, especially in summer when fully grown. The extract must be boiled for about 30 minutes after soaking the plants for 24 hours. It is then mixed in a ratio of 1:5 with water before being applied to the bamboos. It smells very unpleasant so it is preferable where possible simply to mulch with Horsetail, although make sure it is dead first! Bentonite, a silica-rich clay mineral, is very good for supplying bamboo with this important nutrient. It is spread on the soil and washed down by the rain.

If you do not want to be bothered with organic fertilizers, commercial lawn or garden fertilizers can be used instead. Apply 150 g (5 oz) mineral fertilizer each year to every one square metre (square yard) of soil. Be careful not to allow salt accumulation in the soil. Bamboo is very sensitive to salt in the soil and if too much fertilizer is applied this is a danger. The soil should be inspected regularly under valuable bamboo stands. There are mineral fertilizers with reduced salt available. Another useful fertilizer is this: mix 4 parts ammonium sulphate, 1 part superphosphate and 1 part potassium sulphate. Apply this mixture three times at two-week intervals before the start of the growth season at the end of April until the end of May, as above, 150 g/sq.m. (5 oz/yd^2).

Pruning

Bamboos do not need regular pruning. However, wild, unchecked growth is not necessarily best. The culms can live to be eight to ten years old but begin to look unsightly after the fifth year. It is a matter of personal taste whether one allows dense growth or thins them out. In Asia, where they are grown in gardens or temples, they are always carefully thinned so that the full beauty of each individual culm is revealed.

In European gardens and botanic gardens they are usually left to grow more naturally into thick bushes. However, even a dense growth should not be left entirely alone. In the case of old plants the dead culms at least should be removed every year. However, it is better still to remove the four- to five-year-old culms. Five-year-old culms are not as attractive as two-year-old ones because with increasing age the new twigs and leaves that every culm produces each year get shorter and smaller. The young, attractive culms do not develop their full potential if they have to grow up amongst the older ones. The characteristic elegance of a bamboo is achieved if only the best developed

Bamboo avenue in Sotchi, China

culms are allowed to remain. This is particularly important for individual plants or those grown in containers, but even bamboo hedges should not be left entirely untended. Every spring the un-attractive culms should be removed, even if they are large and thick. These can be used for many purposes around the house and garden (see p. 114). Old culms should be cut off at soil level. In China they are split to speed up decay, to make room for new growth, and to supply humus.

In the case of a bamboo hedge there is another reason for cutting out the old growth. A bamboo hedge should provide a screen and windbreak and therefore needs to be thick. Since relatively large species such as *Sinarundinaria*, *Phyllostachys* and *Semiarundinaria* are used for hedges, over the years they develop thick, leafy crowns but the culms below are relatively free of foliage. They therefore begin to lose their effectiveness as a shield, and a tall hedge will start to cast too large a shadow in a small garden. A thicker, not quite so tall hedge is better for most gardens. This is not arrived at by conventional pruning such as might be used for, say, a privet hedge. With bamboo all the older growth must be removed every year. This

reduces the leaf area of the plant and in turn the photosynthetic build-up of new plant tissue. The plant develops only medium-sized stems and its growth is naturally checked.

There is another rather more work-intensive way to keep a bamboo hedge short and thick. By pruning individual culms it is possible to get them to put out more branches and leaves. The culm should only be cut in its upper third after it has developed all its branches. If it is cut earlier it may die back. One disadvantage of this method is that the growth character of the bamboo is lost and one simply creates a thick green wall, lacking any particular beauty. Before pruning a bamboo hedge in such drastic fashion one should wait to see how the plant develops. For example, if you have enough patience a *Sinarundinaria* hedge may provide enough shelter, even though the leaves develop mostly at the top of the culms. These

Growth of a bamboo with leptomorphic (invasive) rhizome. In the first year it sends up a culm, in the second year it grows further, sending up more culms. After the third year the rhizome branches more frequently and grows several metres, now nourished by a mass of leaves

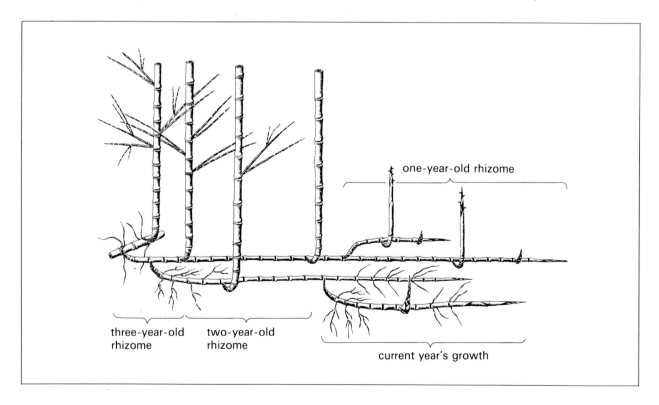

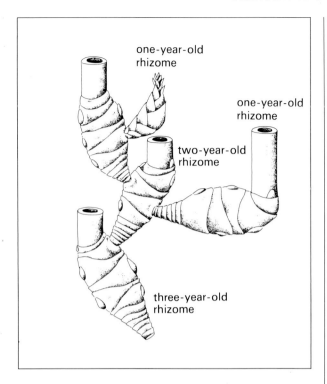

one-year-old
rhizome

one-year-old
rhizome

two-year-old
rhizome

three-year-old
rhizome

Growth of a non-invasive bamboo with pachymorphic rhizome. The thick short rhizomes result in a clumped growth

species bend over and therefore give extra thickness to the hedge. It would therefore be a great shame to prune a *Sinarundinaria* hedge drastically before it is at least ten years old. *Sinarundinaria* occasionally grows too thick as an individual plant and loses its attractive shape. Then you need to take more radical steps and remove all culms older than two years, very difficult in the case of a large, thick plant.

In emergencies, for example when the stems and leaves have been killed by a hard winter, bamboos must be pruned right back. If the leaves freeze or are killed by drought, as in the cold European winters of 1978/9 and 1984/5, but the culms and buds are not damaged, new twigs and leaves will appear. Bamboos should therefore not automatically be pruned back to the ground if the culms have lost all their foliage. There is always the chance that new growth will appear.

If a bamboo is cut right back it will send up long, thick culms the next year, using the reserves from the previous year's growth. But in the following year the rhizome is dependent on food manufactured and transferred to it by the year-old culms. It can take years for it to develop thick, strong culms again. In Asia it takes up to eight years for bamboo groves that have been chopped down to develop strong culms again.

Bamboo species planted as ground cover may be cut back every year or every other year, in spring before the main growth. This can be done with a lawn mower or scythe. Such cutting is good for the plants, even if it does seem rather drastic!

Controlling growth

Some leptomorph species can take over in the garden, especially low-growing and medium-sized species of *Sasa*, *Sasaella* and *Pleioblastus*. These genera have one thing in common: they are extremely active below ground. Their rhizomes grow quickly and become very long. The plants spread out rapidly, often into parts of the garden where they are not welcome. The larger leptomorph species also tend to 'run away' when they have reached a certain age and if the soil conditions are suitable. The tips of the new rhizomes are so hard that they bore through almost everything that gets in their way; they will pierce a pond lining if it is not at least 0.5 mm ($\frac{1}{5}$ in.) thick. Wooden barriers that have begun to rot after a few years will not stop a vigorous bamboo, and even rubble foundations are not safe from the power of the rhizomes, which may grow through into the cellar. However, they do not destabilize the walls like tree-roots sometimes do because they do not thicken. Many garden plants suffer when a bamboo grows along underneath them and takes away nourishment and water. Bamboos are not a problem if one takes certain precautions. Consider, even before you plant a bamboo, especially a leptomorph species, where it can grow unchecked without being a nuisance. Take into account that you need to be able to dig down easily to reach the rhizomes if necessary. For example, a ground covering bamboo should

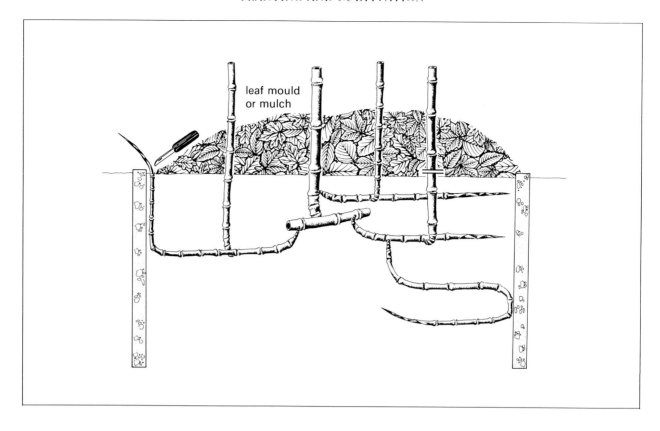

leaf mould
or mulch

Restrict the growth of an invasive bamboo, by using a solid underground barrier. Rhizomes growing out at the edge should be cut off at soil level, together with the older culms

not be sited next to a flower or herbaceous border because the rhizomes cannot be reached without damaging the other plants. On the other hand if it adjoins a meadow, lawn or path it is easy to dig up if necessary. One can use other methods to avoid having to dig up the rhizomes. For example, bamboos may be held in check by concrete barriers. These, however, must be up to one metre (three feet) deep, especially for the low, quick-growing species.

An alternative to concrete is thick plastic. This can be curved to give a natural border between the bamboo and other beds. Bamboos can be completely surrounded by this material. If the barrier is buried 60–100 cm 2–3 ft) deep in the soil, and lawn or flowers grown on top it is completely hidden. The joins, however, must be tight otherwise the rhizomes will penetrate the gaps.

If you want to have bamboo in a flower bed or between other plants, plant it in a large pot or in a plastic barrel sunk into the soil. Make sure that you have a good water supply and effective drainage beneath the barrel. Tall leptomorph genera like *Phyllostachys*, *Semiarundinaria*, *Pseudosasa*, *Bashania* or *Brachystachyum* do not usually send out rhizomes deeper than about 50 cm (20 in.). Again, barriers made of plastic are useful, as are concrete rings which can be purchased at building suppliers. All barriers should be at least a metre (three feet) in diameter, otherwise there is the danger that the plants will dry out too quickly as they no longer have long rhizomes bringing in supplies from further afield. Another possibility is to have an undergrowth of robust perennials or shrubs, but make sure they do not get the upper hand. Both the undergrowth and bamboos must be regularly thinned and pruned to prevent them growing into each other.

If you already have bamboos in your garden with no mechanical barriers they have to be kept in check by other means. This can be done above the ground. The new shoots can be cut off by mowing early in the year, at least once a week, otherwise new shoots will appear for several weeks. (In the case of *Phyllostachys* this is a real shame, because the shoots are a delicacy.) Alternatively, the shoots can be cut off at the rhizome with a sharp knife or secateurs. Destroying the shoots only improves the appearance, however, and the rhizomes continue to grow underground. These have to be capped when they grow in the wrong directions. With the larger-growing genera, whose rhizomes are not so deep, this is relatively simple; they can be severed with sharp scissors and do not have to be dug up. If it is a young rhizome it will die, since it is dependent upon the main plant. If it has been spreading for several years the rhizome will have to be dug out. If you are careful to leave a few culms on the rhizome it can be transplanted to another part of the garden, or given to a neighbour. With the stronger-growing short species it is more difficult to get rid of the rhizomes. They have to be completely removed. If you have to dig them out it is a good idea to put in some kind of barrier, so you do not have to repeat this unpleasant task.

Bamboos in winter

Bamboos are particularly attractive in winter. When the whole garden is grey and colourless and snow covers the lawns and shrubs, a bamboo, with its green leaves and delicate shape, is a wonderful sight. In summer they do not stand out as well in a luxuriant garden unless planted in an open site. But in winter they often stand out as the only green plants with their elegance fully developed. Many species are adapted to snow in their native lands and therefore do well in temperate climates, provided the winters are snowy and without long periods of frost. In such 'normal' winters bamboos are easy plants with few problems. Many species can tolerate several days with temperatures below zero without sustaining external damage – some even to $-25°$ C $(-13°$F).

Longer frost periods are problematic, however, not so much because of the cold but because the plants tend to dry out. Evergreens such as holly, ivy, privet, etc. tend to have thick leaves in which they can store water, or which are protected by a layer of wax, but bamboos, with their thin leaves, are very susceptible to drying. It is also cold in the native countries of some bamboos – one thinks of the Himalayas for example. But there is rather more snow there and in these areas bamboos often grow in a shrub like form.

During extended frost the moisture is literally sucked out of the leaves. Many hardy species protect themselves from frost-induced cell damage by tesselation (see p. 40). If the frost persists the leaves themselves dry off and fall. It is worst when there is a short warm spell between frosts, or if the sun shines very strongly during a frost. The sun quickly warms up the frozen leaves and stems, the cell contents begin to flow and the cell walls burst. The moisture in the leaves evaporates away in the warmth, but the whole above-ground parts of the plant, the stems, leaves and also the buds, dry out because the plant cannot take in water from the frozen soil.

In this case only intervention by the gardener can help, even by very modest assistance. Mulch well, ideally with leaves or straw, in the autumn to prevent the rhizomes from freezing in the soil, even with the hardiest species. The soil should only freeze at the surface and the plants will still be able to take in water from the lower layers. On frost-free winter days it sometimes helps to water bamboos, if the soil is dry. If you are lucky there may be a snow fall before a frost. The snow protects both soil and leaves from the frost. The culms often bend under the weight of the snow to the point where they lie on the ground. Do not make the mistake of shaking the snow off. No harm will come to the culms, which are so elastic that they spring back after the snow has melted as if nothing had happened. It is tempting to knock the snow off the delicate bamboo plants but this should be resisted. The snow protects the leaves from drying out and freezing even if the plant is not bent over to the ground. Apart from that, bamboos look very good with a covering of

Sinarundinaria murielae, green through the snow

as they have not been weakened by excessive applications of fertilizer in summer and autumn. Even when the culms and leaves are completely frozen strong plants will retain sufficient reserves in their rhizomes to enable new culms to be produced. It has recently been shown that bamboos become deciduous in really hard winters, dropping all their leaves and growing new ones in the spring.

Usually, not all the culms will be killed by the frost, even though a bamboo plant may look pale after a hard winter, and its leaves yellow. The buds from which new branches and leaves grow are more resistant than the leaves and it is astonishing to see how a bamboo that seemed moribund starts to put out fresh growth in May, both from the dead-looking culms and from the rhizomes. It is not a good idea to cut back the culms simply to improve the appearance of a bamboo that has overwintered, and sustained drought damage. If this is done, young plants in particular will regenerate only very poorly in the following year (see p. 89). Some gardeners remove the dry leaves after a hard winter. These straw-yellow leaves are attached quite firmly to their sheaths and removing them rips away the sheath as well so that the buds are no longer protected from drying out. It is best, therefore, to leave all the leaves in place on the culms and branches until the new branches have started to grow and the rhizome has begun to sprout. It is also important to make sure the plant has sufficient water, even when it seems to be completely lifeless. Do not cut out the dead culms until the plant has begun its new growth.

In a mild winter with short frosts and snow, sufficient humidity and precipitation, there is no need to do anything to the bamboos, just enjoy them, as long as the soil is not frozen.

snow. However, if it freezes for a long time there is not much one can do to help.

Do not cover the plant with a plastic sheet. Firstly, plastic sheeting is too thin to protect the plant from hard frosts, and secondly, when the sun shines, the heat produced would be even more damaging than the frost.

Valuable bamboo plants or newly planted ones may be covered for a very few days in blistered plastic sheeting, but only when the winter sun is not shining. Experiments using a fleece tent as a snow substitute have proved successful in recent winters.

It is best, however, to let nature take its course. As long as the plants have been well mulched the rhizomes will not be damaged. The culms and leaves can tolerate a great deal of frost as long

Propagation

There are two ways to propagate temperate bamboos: generative propagation with seeds, and vegetative propagation using sections of rhizome, with or without culms.

Seeds are not often available in Europe and collectors usually get these from overseas. However, when a bamboo flowers and produces seeds it can sometimes regenerate via the seeds. If you have a flowering bamboo (*Pseudosasa japonica* has flowered in our garden for several years) you can raise seedlings from it, and this is an interesting process. After ripening, put the seed in a pot with normal garden soil, mixed with a little peat. The seeds can be stored for only a couple of weeks before they lose their viability. Cover the pot with glass or plastic. This promotes better germination by maintaining an even temperature and humidity. With luck the seeds will sprout in two to six weeks, producing a tiny bamboo shoot that looks just like a grass stem. Keep it warm and damp, ideally indoors or in a greenhouse. Be very careful when watering because the tiny roots rot easily.

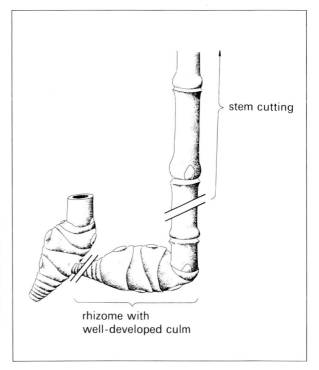

Propagating a bamboo with pachymorphic rhizome. Cut off a section of rhizome that is not older than a year, leaving at least one culm. The culm can be cut back to reduce transpiration, but leave a couple of branches. The drawing also shows where to cut a culm section, for horizontal rooting in a warm greenhouse. This works only with tropical bamboos

Young plants of *Pleioblastus pygmaeus* var. *distichus*

It takes about three years to grow in to a small bamboo plant that is ready for planting out.

For most people vegetative propagation, using rhizomes with culms or by dividing a large plant, is the easiest option. Bamboos should be divided if they are too big or if you want another plant of a particularly attractive species elsewhere in the garden, or as a present for another bamboo enthusiast. Dividing a large bamboo is quite hard work. It has to be cut into two or three pieces with a sharp spade, although sometimes an axe or saw is necessary. The pieces should be planted in the new site with all their culms and as much earth on the rhizomes as possible. When it is safely planted remove all old and unsightly growth so that the rhizome does not use up reserves on such old growth, but uses these instead for producing fresh young culms. The best time to divide a

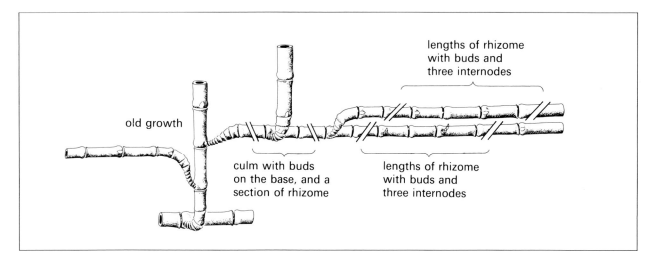

old growth

culm with buds
on the base, and a
section of rhizome

lengths of rhizome
with buds and
three internodes

lengths of rhizome
with buds and
three internodes

Propagating a bamboo with leptomorphic rhizome.
Cut off a piece of two-year-old rhizome, at least 30 cm
(12 in.) long, with one or more culms and several nodes.
Alternatively, use a fairly long section without culms, but
with at least three nodes with clearly visible buds

bamboo is before the new growth appears at the surface – for most hardy genera is early spring (see p. 32). Tropical bamboo genera are best divided in late summer, which is when they put out new culms – August or September, depending on the weather. *Sasa* species recover best if they are not divided until after the first leaves have begun to unfold. *Sasa* and *Pleioblastus* are the easiest to divide. It is quite wrong to divide bamboos in late autumn, in cold climates. The divided rhizomes, which suffer some inevitable damage, find it difficult to recover and the plants will take two or three years even to retain the growth they had before division. Division in early spring is therefore best, because in hardy species the reserves are then spread throughout the whole rhizome. The buds are about to sprout and the roots are ready to develop. The plant is activated at this time of year and primed ready for the growth of new rhizomes, culms and roots, so division is not so damaging to its development.

Another form of division is not to cut up the entire plant, but to grow a young plant from just a piece of rhizome. You must select the right time of year for this operation too. The best time is during the development of the buds in the rhi-

zome, before the new culms grow. This stage is at different times of year according to the species, the climate and nourishment. One cannot therefore generalize, but each gardener must observe the individual plants carefully to establish the right moment. Rhizome pieces will only send out new culms if they contain active young tissue.

Clump forming bamboo species with pachymorphic rhizomes are propagated in the following way. Part of the plant is cut off at the rhizome neck, the narrow point of attachment. Such a wound rots quickly, but only as far as the active tissue. Such bamboos can only be propagated using rhizomes that are up to a year old, since only these can send out the side rhizomes which are essential for further development. The year-old rhizomes can be identified by the young roots they carry. Older rhizomes are not able to grow new rhizome. On the young rhizome there must be a culm, with branches and leaves, although evaporation can be reduced by cutting back to the first few internodes, as long as at least two branches are left on the culm. If you remove the culm completely the rhizome does not develop its roots as successfully.

To propagate species with leptomorphic rhizomes you need a piece of rhizome at least 30 cm (12 in.) long with several nodes, and it should be one or two years old. There is a good reason why

(*Right*) Eighty-year-old stand of *Phyllostachys viridiglaucescens* (Schlossgarten of the Markgraf, Baden-Baden, Germany)

one should not use much older rhizome pieces, and this is connected with the development of rhizome and culm. In the first year the rhizome grows out horizontally, growing a bud at each node. In the second year the rhizome continues to grow horizontally, with new bamboo growth at the buds. By using one- or two-year-old rhizomes one ensures that the buds are already developed so that one does not need to wait too long before culms appear above ground. It is best of all to use a piece of rhizome that already has one or two culms on it, as well as rhizome growth from the previous year and a piece of rhizome should have buds clearly visible on at least three of its nodes. Within a few weeks of planting one or two culms should be visible, depending on the amount of reserves stored in the rhizome. By using a piece of rhizome with one or more culms you can reduce the propagation time by a year. On the other hand rhizomes with culms, branches and leaves need more feeding and watering than a culmless rhizome. If it is not looked after well such a rhizome can be so exhausted by the culm that it has difficulty producing culms, buds and fresh rhizome in the following year.

If you plant the piece of rhizome vertically in a pot it will begin by sending out culms, but if planted horizontally the roots will develop first. The plant is not ready for planting out until the new culms have grown their own roots and the buds of the basal nodes have developed fresh rhizome.

A plant propagated from a piece of rhizome in the spring of year one is ready to be planted out in the spring of year three. Until then it is best to keep the young plants in plastic pots in a warm, bright position.

Sinarundinaria nitida and *Chusquea couleou* have pachymorphic rhizomes and must be propagated using pieces with as many shoots as possible and at least one culm under two years old. In their native countries tropical bamboo species are propagated by other methods, using the culms. In temperate climates these methods work only in greenhouses. A two-year-old culm is cut off at soil level and placed in a long furrow in the soil about 15 cm (6 in.) deep, covered with earth and kept warm. Still air under a plastic cover encourages development. After a few weeks the buds sprout roots and culms, and after a few months the plants are separated by cutting through the internodes and transplanted some time later. In Asia bamboo groves are rejuvenated in the following way. A thick bamboo culm is sawn into short sections, if possible just two internodes long, leaving half an internode at each side of the node. These are filled with wet sand and the sections placed in warm soil where they take root and send up fresh culms. Unfortunately these simple methods are not suitable for a temperate climate, nor for the hardy bamboo species. But if you have a tropical bamboo species in a conservatory, and green fingers, it is always worth a try.

7

Garden Design

Bamboos are becoming more and more popular every year, and for very good reasons. The main reasons are definitely their transparent beauty and elegance as well as the ease with which they can be grown. They are also highly prized because, as evergreens, they look attractive in gardens even in the depths of winter. Bamboos also do not compete with other plants visually – they have such a delicate structure that other plants look good when grown near them and may even be highlighted. Bamboos can also be used in a variety of ways in the garden. In recent years such a range of genera and species has become available that one can buy bamboos in all shapes and sizes and for many different situations. They can be planted as a hedge, as a grove, as ground cover, or as solitary plants. Bamboos love courtyards and balconies, in the open as well as on roofs and even window boxes.

As far as we know bamboo is not sensitive to pollution and remains healthy when native plants succumb to the effects of acid rain. There have even been plans to replace forest trees with bamboo.

If you are buying a bamboo for your garden, courtyard or terrace you will have a pretty clear idea of the exact conditions of the site, which must be suited to the particular requirements of the species. Here we provide suggestions only as to which species are suitable in different situations.

Individual plants

In Japanese gardens one often sees just a few bamboo culms in gravel, which is raked in a particular pattern. Whilst this may look beautiful it is not our idea of a garden. However, one should not forget this basic idea completely when considering bamboos as individual plants, and bamboos should be given sufficient room to display themselves to their full potential. An individual bamboo certainly does not always look right in front of a green hedge or shrub. Part of the beauty of a bamboo comes from its elegant shape, part from the delicate leaves and from the green or coloured stems, and all of these features should be clearly visible in a solitary plant. Bamboos therefore often need a neutral background such as a white or red brick wall. A tall wooden fence can provide as effective a backdrop as dark conifers. The beauty of an individual bamboo standing in the middle of a lawn can be emphasized by a simple fountain, a Japanese stone lantern or an interesting rock. Other shrubs should be kept away – the bamboo is the star and should be allowed to shine alone. An individual bamboo is most effective near to a garden pond or small stream. It is reflected in the water, thus doubling its beauty. However if it is surrounded by the usual tall waterside grasses its beauty will be hidden. Waterside bamboos go better with stones and low, colourful perennials.

The taller the bamboo the nearer the surrounding plants can grow without destroying the effect. A 5 m (16 ft) bamboo looks good amongst other shrubs which would rather smother one only a couple of metres tall.

The best bamboo for individual plants are usually *Phyllostachys* species, for example *P. decora* or *P. nigra*, and *Sinarundinaria*, such as *S. nitida*, 'Eisenach' or 'Nymphenburg', *Pseudosasa japonica* and *Sasa palmata*. Those with a more delicate growth, such as all *Sinarundinaria* species, are best for the smaller garden. A small garden with a well-grown bamboo in the middle of a bed of colourful annuals and perennials seems larger than it would with a large conifer or thick shrub instead. *Phyllostachys* species with their straight, upright growth are very good for narrow gardens.

(*Above*)*Phyllostachys nigra* and
P. flexuosa amongst stones on an
artificial stream

(*Right*) Green roof formed by
Phyllostachys aurea over a natural
stream

(*Above*) *Sinarundinaria murielae* next to a courtyard fountain

(*Left*) Hedge of *Sinarundinaria nitida*, underplanted with busy lizzie (*Impatiens*)

Hedges

A bamboo hedge is both attractive and useful. It provides a screen and windbreak for the garden and is easy to maintain, and because it is evergreen it has a number of advantages over most hedges. It is attractive throughout the year and continues to provide the garden or terrace with protection from the wind when it is most needed. A bamboo hedge creates a favourable microclimate from which all other plants will profit. In England *Pseudosasa japonica* is particularly common as a hedge plant. It grows very thick and because it has relatively broad leaves it is very effective as a windbreak. In France *Semiarundinaria fastuosa* is commoner. This is planted between streets and houses because it provides a barrier against noise and dirt. Such hedges are very straight and strong but have their own appeal. *Sinarundinaria* can also make a delightful hedge and in this case it is not just the leaves, but also the many culms, that

The older the bamboo culm, the more it branches, and the more leaves it carries. This is useful to know when growing a hedge

provide the protection. Birds like to nest in bamboo hedges and use fallen leaves and sheaths for nest-building; cats find it very difficult to clamber up the culms to reach the nests.

The older the hedge the denser it becomes, and the closer together the young bamboos were planted the quicker it thickens up. Planting a new hedge with *Phyllostachys* at 1.5 m (5 ft) intervals will give an almost closed hedge, provided the planting and maintenance are carried out correctly. With some species there is a danger that after a few years the hedge will remain dense only in the upper part and this is true for *Phyllostachys* species. This is because the taller culms tend to put out branches and leaves at the tip. It is therefore best to keep a bamboo hedge relatively short to maintain its screen and windbreak qualities. This is not achieved by conventional hedge pruning and the hedges would soon become unsightly if all the stems were cut back to the same height. Each culm would put out more branches but the hedge would not look very good and would be quite untypical. Another method has to be used whereby the hedge is continually thinned. A proportion of the older, more unsightly culms are cut off regularly at ground level. The

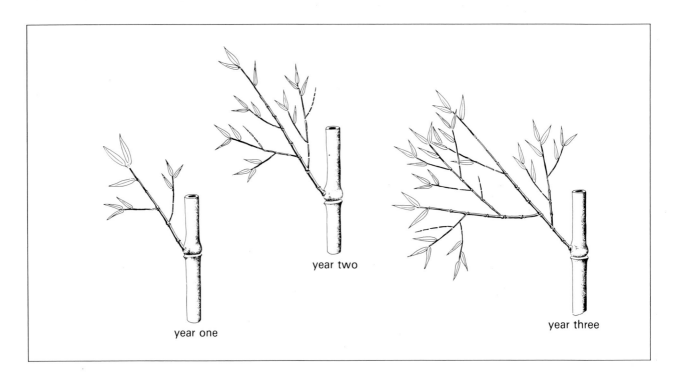

year one

year two

year three

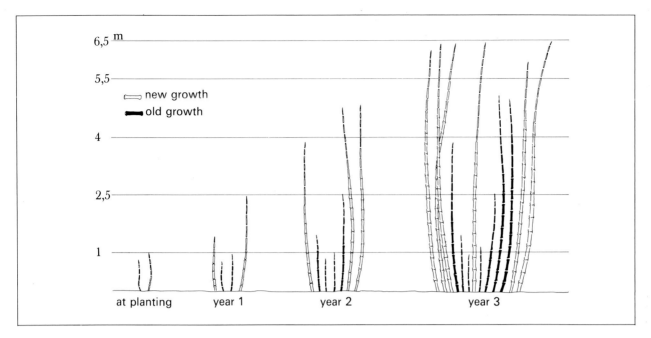

6,5 m

5,5

▭ new growth
▬ old growth

4

2,5

1

at planting   year 1   year 2   year 3

Development of a young bamboo plant under ideal conditions. The young culms are taller and thicker than those of preceding years. Evenly sized culms in subsequent years are not produced until the plant is five to ten years old

hedge will only keep its shape if it is regularly pruned in this way.

Monopodial species can grow very long rhizomes. If you do not want a hedge to turn into a grove the shoots that grow up outside the boundary of the hedge must be dug up (and can be eaten in the case of *Phyllostachys*). Alternatively they can be removed with a lawn mower. For the neighbours' sake chop off the rhizomes with a sharp spade, or keep them in check with stone blocks or plastic, sunk at least 60 cm (2 ft) into the soil.

Bamboo groves

If you have a large garden you can allow a group of bamboos to grow into a grove and this is one of the most fascinating ways to grow bamboos. All the taller species are suitable – especially *Phyllostachys* and *Semiarundinaria*, *Phyllostachys* being the most attractive. *P. aureosulcata* and *P. bissetii*,

for example, grow very upright, but the arching species also look good.

The so-called 'giant bamboo', *Phyllostachys viridis*, is also suitable. In temperate latitudes this species does not grow as huge as in its native land, but it does reach 20 m (65 ft) in southern France and Italy, forming some of the most impressive bamboo groves in Europe. In areas that are not too cold *P. decora*, *P. rubromarginata* or *P. viridiglaucescens* are good species to plant.

One has to be very patient in planning a bamboo grove because it takes about ten years to reach its full height. First put three or more plants in position, making sure there are no other trees or shrubs in the vicinity. While the grove is growing the area can be made more attractive by placing beautiful stones amongst the bamboos, and sowing annuals and biennials in between. Bushes and shrubs are not suitable because the large bamboos would compete with these and take away their light, nourishment and water.

The bamboo plants grow more and more powerfully each year, until they reach their maximum height after about ten years. The taller the plant, the more strongly its rhizome spreads and the thicker its culms become. The more culms, branches and leaves it grows, the better will be its supply of nutrients to the rhizome, allowing it to store more reserves, in turn sending out stronger culms. The bamboo grove develops tall culms with a dense leafy roof – just the features one strives to avoid in a bamboo hedge. However, the grove should be thinned every year. It is sufficient to remove the four- to five-year-old culms and for this you can use the method favoured by professional bamboo planters in Asia. Mark the culms in their first year of growth using a permanent marker pen. In this way it is always clear which culms are ready for removal. If you are familiar with bamboos you will know from its appearance and amount of branching whether a culm is ready for cutting or whether it should be left for another year.

When the grove is fully grown you have it, so to speak, for life. Be careful therefore that it does not take over the rest of the garden. It will grow steadily larger and take away light and nutrients from other garden plants, and for this reason the siting of a bamboo grove needs very careful thought indeed.

Ground cover

The low-growing bamboo genera such as *Sasaella* and *Pleioblastus* are useful for providing ground cover and indeed are often planted like lawns in Japan. *Pleioblastus pygmaeus* is particularly suitable, as are *P. pumilis* and *Sasaella ramosa*. Although these species grow to 80 cm (30 in.) they can be kept shorter and thicker by occasional mowing. *Sasa* and *Sasamorpha* do not spread as quickly but are good for planting under trees and large shrubs because they are shade-tolerant. *Sasa veitchii* is very attractive with its leaves that dry from the edges in autumn.

To make a quick ground cover of low bamboos you have to plant 8 to 12 per square metre (square yard) in the spring, which is expensive. It is cheaper to start with 1 to 4 plants per square metre (square yard) and to sow summer-flowering annuals between them. After about five years the leaves will have closed the gaps, depending on the spreading capacity of the bamboo and the soil conditions.

The following rule of thumb for planting ground covering bamboos may be helpful. However, bought plants do vary a lot in size. To cover 1 square metre (square yard) it takes:

1 plant	4 to 5 years
3 plants	3 to 4 years
5–7 plants	2 to 3 years
8–12 plants	1 to 2 years

The spread may be faster in warmer climates.

Ground covering bamboos are perfect under large trees or thick shrubs. They prevent weeds becoming established and also stop the soil from drying out. They look particularly good beneath large conifers where the dark and light greens make an attractive contrast. They also look good under lilac and rose bushes as well as beneath natural hedges, where they keep the weeds at bay. *Sasa* species are especially suitable under trees and hedges because they thrive in the semi-shade.

Bamboos are ideal for stabilizing steep slopes in the garden, although they must have sufficient water, which is not always easily supplied in such sites. After only a few years the rhizomes are so firmly matted together that the soil is held in place even in the fiercest rainstorms. This is particularly important in gardens that have been established in new housing developments. Earth moving during the construction period thins the soil in steeper places, but in the early years deep-rooting plants that would hold the earth in place tend not to colonize. In such circumstances bamboos can be a real help. They can do well even on south- and west-facing slopes, as long as you do not expect a very vigorous growth.

Ground covering plants, particularly *Pleioblastus* species, need careful planning. They can run away in small gardens and quickly become a nuisance because they rapidly send out long rhizomes and always seem to turn up where they are not wanted. When they spread out rapidly they can also damage other plants. All plants used

(Above right) *Pleioblastus pumilus* with summer marguerites

(Below right) *Sasaella ramosa* used as ground cover and path edging

(*Above*) *Phyllostachys viridis* by a lake

(*Left*) *Pleioblastus pumilus* with summer perennials

in conjunction with ground covering bamboos should therefore be taller so that they can hold their own against the bamboos.

Bamboos in the open

Groves, hedges and 'meadows' of bamboo, as well as large individual plants, are particularly suitable for open places. Bamboos require little attention and are relatively insensitive to air pollution. In Japan for example, where pollution was a big problem before the introduction of catalytic convertors, bamboos suffered no obvious damage; they seem to be sensitive only to salt in the soil. Bamboos are quite at home in built-up areas and some cities have made use of this. For example, Bern University in Switzerland has large bamboo plantations lining the roads. Bamboos are increasingly planted around public buildings. Their light-green foliage ameliorates drab concrete, even in winter, and bamboos are also seen in public parks and occasionally in pedestrian precincts. Europeans and Americans are just beginning to make use of bamboo in the ways that the Japanese have done for years, particularly as street planting. The thick rhizome layer holds the plants firmly in place and there is no danger of heavy rain washing soil out on to the streets. In Japan such bamboo beds are mown every year, promoting a thick, short growth.

Bamboos on the terrace

A terrace is an excellent site for growing bamboos. Bamboo provides a splendid loose shield throughout the year for terraces in built up areas. Behind a bamboo screen one does not feel hemmed in, since the bamboo is effectively a light-green curtain which lends a special atmosphere to the terrace. The patterns of light and shade produced by the leaves, even in the slightest breeze, have moved the Chinese and Japanese for centuries to regard bamboos as 'house art'. In Asia people are capable of meditating for hours on the shadow produced by a bamboo branch.

Species with delicate foliage and attractive branches such as *Phyllostachys aureosulcata*, *P. bissetii* and *P. humilis* are particularly good for terraces. These species are all beautifully branched. The branches of *Sinarundinaria nitida* 'Eisenach' and 'Nymphenburg' hang down like green cascades. *Phyllostachys bambusoides* 'Castillonis' has a yellow culm with a green sulcus and occasionally white-striped leaves and looks very fine near the house, although it is not quite as hardy as other *Phyllostachys*. Growing near the house helps to protect the plant from extreme cold, strong winter sun and sharp winds.

Phyllostachys aurea grown as a screen

Old grove of *Phyllostachys nigra* f. *henonis*

Bamboos on the terrace are particularly welcome in winter. On the greyest of days they provide a touch of spring, and when the snow clings to the green leaves and branches, bending the culms right over, the sight is a source of pleasure and wonder.

Terrace bamboos can be underplanted with flowers. However these must be species that do not need much light. Ferns can also look attractive beneath bamboo that is not too thick and busy lizzie (*Impatiens*) in all colours grows well and look splendid. Low-growing roses are another possibility. These go well with delicate bamboo foliage, even though they may not flower as profusely as they would in full sun.

Bamboos in the courtyard

Bamboos are particularly well suited to courtyards, or for covered spaces. The rhizomes seldom damage the flooring and if a shoot grows up from between the stones it can be removed easily, or left to grow. This can look extremely attractive. Bamboos like the protected climate of such areas and it is possible to grow species that would not flourish outdoors. However, do not forget that quite sharp frosts are possible even in courtyards.

Arrangements with bamboos in courtyards look especially attractive if the bamboo is given central place in the display, possibly together with a pool and large stone, or with ferns, mosses and ivy planted underneath. If you do not want the courtyard plastered or concreted you could try a Japanese display, with bamboo and white gravel raked with graphic designs. Such a yard is purely decorative and cannot be used for general outdoor activities. When siting bamboos in a courtyard remove some of the floor stones so that the bamboos can be planted direct into the soil. If they are planted in a trough this must be protected from frost in winter because the earth might otherwise freeze solid. Courtyard bamboos need to be watered more frequently than those outside. This is partly because the floor covering prevents the rain from soaking into the soil and partly because the walls and overhangs provide some protection from the rain. Tropical species still need to be brought indoors in winter, to a bright position (see p. 109).

Roof sites

Bamboos are beginning to be used to great effect on roofs, sites that are not normally accessible. Although green roofs cannot replace parks and gardens in built-up areas, flat roofs provide much potential space for oxygen-producing plants which can improve the climate in inner city areas. They also give flats and offices some protection from extremes of climate. In summer the plant and soil layers prevent excessive heating of the roof and the air underneath, and in winter they help to keep out the cold.

Before planting up a flat roof it has first to be covered with a 1 mm thick sheet to contain the root growth and this must be of a material that will not rot. This prevents the roots from penetrating the roof and making it leak. A nylon mat ensures the necessary aeration of the roots and also lets excess water run out, whilst a fleece mat on top of this prevents any of the substrate from escaping. Since soil is often too heavy to use, especially if the building has not been designed with a roof garden in mind, stone-wool mats are often used as a substrate. These are lightweight and easily transported to the roof. However they are only available up to 20 cm (8 in.) thick, insufficient for bamboos and shrubs, let alone small trees. Bamboos are therefore planted in special tree baskets that prevent the rhizomes from spreading out and penetrating the roof. Thus one cannot completely cover a roof with bamboos, like a lawn. The rhizomes would spread out too strongly and would sooner or later break through the protective sheet. It is also too hot, dry and windy for a bamboo 'meadow' on a roof site. They should therefore only be used in conjunction with other plants and in sites that are accessible for watering and fertilizing.

Only reliably hardy bamboos, and other species that do not grow too large should be planted on roof sites. Trials with *Sinarundinaria murielae, S. nitida, Pseudosasa japonica, Pleioblastus pumilus* and other *Pleioblastus* species have proved successful.

Roof gardens

As long as they are sufficiently strong, roofs provide a good site for creating gardens. The same applies here as for courtyards and terraces. Only reliably hardy species should be used, as recommended in the previous section. It is commoner however to use containers on roof gardens and the following chapter should therefore be consulted as well.

8

Growing Bamboos in Containers

Planting and cultivation

Even if you have no garden or live in a city flat you can still grow bamboos because they make very good container plants. In theory one can grow almost all bamboos in containers, on a terrace, balcony, roof garden, conservatory or at a bright window inside the house. Tropical species are often used, since these can be brought inside to avoid the hard winter weather. One uses different criteria when considering container plants. Garden plants need to blend in well with others to make the overall picture, but in the case of container plants it is the individual beauty of the plant that is of prime importance. Whenever possible use species that have a particularly beautiful shape or special features such as interesting stem colours, beautiful sheath patterns or elegant foliage. All *Arundinaria* species, *Bambusa* and some *Phyllostachys* species, such as *P. flexuosa*, *P. nidularia* or *P. nigra* are particularly attractive in containers. The drooping forms of *Sinarundinaria* are also very suitable.

As in the case of garden bamboos, container bamboos need the right site for optimal development. Some need full sun whilst others do best in semi-shade, depending on the species and genus. Above all the container must be large enough – less than 50 litres (11 gallons) is far too small for most of the larger species. Unlike outside, where the rhizomes are free to spread, in this case they are restricted. Like other plants, bamboos become stunted if the culms and leaves are not sufficiently well nourished by the roots. A large container is therefore absolutely vital if one wants to grow a magnificent bamboo.

Cutting

If you do not want the bamboo to grow too large in its container or pot – and for house and balcony plants this is desirable – it is possible to keep it small and bushy. The trick is to divide the plant, for several years running, into individual culms, each with just two or three rhizome buds. From these you get fast-growing but small young plants that can be kept in small pots. If you a well-developed bamboo in a container make sure it does not develop too many culms, otherwise the restricted rhizomes will not be able to support them all. Remove a third to a half of the older stems on a regular basis. This should be done every year or every other year, depending on how many fresh culms appear each spring. If only four or five new culms appear in May, just remove the older unattractive culms. However, if there are very many new culms, remove all of the older ones, together with the weaker fresh ones. This ensures that the aerial and underground parts of the plant are in balance and that they will obtain sufficient water and nutrients. Cut the culms as low as possible, but leave the stumps to rot. They will disintigrate within a few months and provide the best humus for the plant.

Watering

It is very important with bamboos in containers not to let the soil dry out. The soil in the container will warm up much more quickly than in the garden and also dry out quicker, and in addition the plant's transpiration rate will be higher. The plants must therefore be watered more frequently than those outside.

With a large bamboo it is a good idea to have a layer of gravel as drainage, so that water does not accumulate and rot the plant. There is a greater risk of this happening than when planted outside.

It is now possible to buy cleverly constructed plant containers with drainage and a water reservoir incorporated. These are deeper than normal containers and the reservoir has a plastic cover penetrated by wicks, usually made of cotton wool. The wicks carry the water to the substrate from the reservoir, which is topped up by a pipe. The advantage is that water is supplied to the plants at the right rate. The disadvantage is that you only notice if the wicks are no longer working properly when the plants are completely dried out. There is another relatively recent development and this is the reservoir-container filled not with soil but with rock-wool matting. The rock-wool can be cut to fit the pot and is particularly useful for bamboos grown as house plants.

Bamboos in pots or containers are best watered with rainwater, especially in hard water areas. Hard water leaches the soil and makes it less permeable. This happens with other plants too but in the case of bamboo it is difficult to renew the soil without damaging the rhizomes. Prevention is better than cure. Attempts with hydro-techniques have tried, but the rhizomes nearly always rot. Nevertheless, they are still trying and although this may become possible in the future the hobby gardener is not likely to have any success with hydro-culture.

Fertilizing

Container bamboos can be fed in the same way as those in the open. But since container plants are brought indoors in the winter one can continue to feed them during the summer. Fertilizer should be applied to container plants when the culms sprout, then three months later when the rhizome develops further and then two or three months after that, when reserves are laid down in the rhizome. For hardy genera this would be in April, June and September, for tropical genera that begin to put out their culms in late summer or autumn, somewhat later. If the plants are left out over the winter, either with insulation or sunk into the soil, no fertilizer should be applied in late summer or autumn.

Overwintering

Overwintering bamboos is somewhat trickier than for other container plants. They need a great deal of light in winter, otherwise their foliage deteriorates. Bamboos will not survive overwintering in a reasonably light cellar or even a garage. They can be left in an unheated room as long as the rhizomes do not freeze inside the pots. A greenhouse kept at about 5°C (41°F) is best for most genera, although tropical genera need at least 10°C (50°F). They can be placed in a conservatory or very bright room (see p. 111). Hardy species may be left outside as long as the container has a thick insulating cover. This might be made of glass fibre or, alternatively, straw or compost. The best solution is to sink the container into the soil and pack with a mulch.

Balconies and roof gardens

Bamboos are very suitable for balconies, as long as they are not too hot and sunny, and for roof gardens, but they need protection from wind and from drying out. A glass wall provides a good windbreak for a balcony. But make sure it is unbreakable and that you stick bird silhouettes on it to prevent birds from flying into it and fatally injuring themselves. A trellis with climbing plants is also a good windbreak. Roof gardens tend in any case to be designed to give protection from breezes.

It is possible to create a jungle-like tropical thicket on a balcony or roof garden using bamboos in containers. But bear in mind that they all have to overwinter somewhere, and this can be quite a problem if they grow larger. Large specimens are best kept in light containers so that it is easier to move them. As long as the container is big enough *Sinarundinaria* can be kept on the balcony throughout the winter in mild areas and its rhizome can withstand some frost.

House plants

Bamboos need a lot of light and this is reduced, compared with outside, even in the brightest rooms. This reduces the choice considerably and the most suitable are those which like to grow in semi-shade, such as *Bambusa glaucescens* 'Golden Goddess', *Drepanostachyum falcatum*, *Pseudosasa japonica*, *Sinarundinaria nitida* and *Indocalamus tesselatus*. *Phyllostachys* also grows well at a sunny window. Do not worry if your bamboo 'cleans' itself, that is it sheds its leaves (and leaf sheaths) frequently, and over a long period. The leaves are easily cleared away and this is a small price to pay for the joy of having a decorative house plant. The amount of light available inside cannot really compare with conditions in the open and even semi-shade species need a lot of light when grown indoors. This can be tested quite simply by using a camera light meter to measure the light intensity outside, at the window and then a short distance away from the window: the result will probably surprise you. Although our eyes can scarcely detect such differences, plants, which require light in order to synthesize food, are very sensitive and react accordingly, especially bamboos. Therefore, always site a bamboo plant in the brightest possible part of the room. Their leaves will not burn, even at sunny south-facing windows, although true shade species will sometimes do better at a west- or east-facing window. If a bamboo is very close to a window it tends to grow its leaves towards the glass to maximize the light and this can spoil its appearance.

Conservatories and greenhouses

Conservatories and glass verandahs are getting more popular. Many people like to live through the year surrounded by green plants, but temperate climes do not allow this. With a conservatory, especially when it is well planted, it is possible to

(Left) Phyllostachys aurea 'Albovariegata' in a large container. The *Campsis* flower is trailing down from above

extend the summer by months. More and more balconies are being turned into conservatories and whole glass façades are often now incorporated into new buildings, primarily to save on heating costs. The sun heats the conservatory, which in turn heats the adjacent room when the doors are opened. In some conservatories the floor and walls are so constructed as to store the heat and release it slowly. In summer, when it is hot outside, and even on sunny winter days, conservatories can overheat. A further problem is condensation, and to avoid both these drawbacks it is essential to have adequate ventilation and also some shade. The conservatory enables tropical bamboos to be grown but these need a high humidity as well as heat, even in summer when the air is usually dry. A water trough will supply the humidity, otherwise a spray is necessary. Tropical bamboos need to be kept in the warm even in winter. If you have a fan heater to ensure the temperature stays above about 10° C (50° F) they can stay in the conservatory. Special greenhouse fan heaters are available for this purpose although a simple heater with a thermostat will do for small spaces.

Bamboos do well in bright, glass-covered courtyards, and one sees them increasingly in public buildings. In the USA whole bamboo groves are sometimes planted in the vast bright entrance halls and foyers of industrial concerns. Large bamboos are also planted in entrance halls and bright courtyards of offices, firms and even hotels.

Bonsai

The art of growing miniature plants and gardens, that is bonsai, has a long tradition in Japan and China and it is only natural that bamboos should be involved. Bamboos are part of everyday life in Asia and their delicate shapes lend themselves to bonsai.

It is not very easy for an amateur to grow a bamboo as bonsai, but it is well worth a try. It requires a feeling for the precise growth of the plant, as well as much care and attention. It is worth the effort because you do not have to wait years for success, as you do with tree bonsai.

Pleioblastus chino var. *gracilis* as bonsai

Bonsai with low-growing species

There are different ways of cultivating bamboo as bonsai. The simplest method is to plant a low-growing species, such as one from the genus *Pleioblastus* in a shallow bonsai bowl, give it reduced fertilizer and water and remove the two-year-old growth. The shallow pot contains relatively little substrate, so the rhizome remains small and is only able to send up small culms.

The water supply has to be regulated quite carefully so the plant neither dries up nor drowns. One feed during the growth period is enough for a bonsai plant and, with a small species, you can maintain a height of about 20 cm (8 in.)

It is more interesting though to use species with unusual growth, coloured leaves, thickened internodes or remarkable culm colours. Examples are *Bambusa ventricosa* (Buddha's belly), in which the internodes thicken up when it is grown under conditions of reduced nourishment, *Phyllostachys aurea*, with shortened internodes, and *Chimonobambusa quadrangularis* which has square culms. Others which are particularly attractive as bonsai are *Bambusa vulgaris* 'Vittata' and *Phyllostachys bambusoides* 'Castillonis', both of which have yellow culms with green stripes, *Chimonobambusa marmorea* with its red culms and patterned sheaths and *C. m.* 'Variegata' with white-striped leaves. In China species with a white bloom on the culms are valued in bonsai. *Pseudosasa japonica*, *Bambusa glaucescens* and its various cultivars, and *Phyllostachys humilis* are also suitable. As yet we are relatively inexperienced in bonsai and until recently the number of species has also been rather limited. This is, however, beginning to change and the choice of species is growing steadily.

Bonsai through culm pruning

In Asia they use more subtle methods to make dwarf forms of normal bamboos.

With sympodial, that is clump forming species with pachymorphic rhizomes, all the culms of the young plant are cut off just before the rhizome's growing season. This is in early summer when the culms have branched and the leaves developed. The rhizome is then planted in a shallow bowl in a mixture of soil and sand. Since the rhizome was unable to store food produced by the previous season's growth the culms produced the following year are smaller than normal and if the procedure is followed the next year the growth gets even smaller, as do the branches and leaves.

For monopodial, invasive species the method is to dig up about 30 cm (12 in.) rhizome as for propagation (see p. 92) and plant this in soil and sand, either vertically, or better still in the shape of a bow, with the curved section (which must have buds) about 6 cm ($2\frac{1}{2}$ in.) under the soil and each end poking out. The buds under the soil

send out small rhizomes and the ends of the rhizome develop small culms with branches and leaves.

Bonsai through peeling the culm

The Japanese bonsai master Koichiro Ueda recommends the following method. When the shoots have grown to about 8 cm (3 in.), the culm sheaths are removed, repeatedly during the period of culm growth, from the bottom upwards. These sheaths contain growth hormone (see p. 32) and when they are removed internode growth is halted.

Removing the sheaths is tricky, and cannot be done by hand since this would damage the soft culm. It is best to use a fine sharp pair of scissors inserted between the culm and sheath. This way the sheaths can be cut into strips, from the tip towards the nodes, and then carefully removed so as not to damage the buds from which the branches and leaves will develop. Start with the lowest internode and then watch the growth of the second, peeling off the sheath at the right moment. This can take up to a day, and must be repeated until the whole culm has been peeled. Bamboos also grow at night and Ueda recommends watching through the night as well, if necessary. The bare culms are very delicate and soft and need to be protected from direct sun until they are firm and have sent out their branches and leaves. Feed well in early summer and early autumn so that the resultant bonsai plant will develop a good colour and do not be too sparing with water in the early stages. In autumn transfer to an attractive bonsai bowl, which will restrict rhizome growth. If the original plant had several culms it is possible to grow a miniature bamboo grove in due course.

Bonsai bamboos look best indoors, as room decoration, but even these mini bamboos should be given a spell outdoors, especially the hardy species, which benefit from a winter rest period.

9

Bamboo as a Raw Material

Bamboo culms removed when hedges are thinned or individual bamboos pruned are much too useful to be simply thrown away. They can be used for many purposes, although not quite as many as in their native land. There the culms are much thicker, but even the thin culms gathered from the garden can be used for all sorts of handicrafts. A single cane is a good support for a tall pot plant, or for staking herbaceous plants in the garden. Several canes can be tied together to make a trellis as a support against a wall for roses and other climbing plants. Make sure that there is a gap of 1–2 cm ($\frac{2}{5}$–$\frac{4}{5}$ in.) between the trellis and the wall so that the plant can twine round it, and also that the hooks holding the trellis are firmly secured – climbing plants can be quite heavy after a while. A bamboo trellis also makes a good partition between balconies, or as a windbreak, with a vine, knotgrass or runner beans. Push the vertical canes into two pieces of wood with holes bored in them, one below and one secured above. The horizontal canes can then be woven between these. It is easier to use freshly cut canes which are still pliable.

Bamboos are also good for fences, although the canes grown in temperate climates are not thick enough for a really stable barrier. However, for decorative fences the thinner canes are ideal. The canes may be pushed into the soil in an arched pattern and tied where they cross over. This is effective in keeping dogs or poultry out of vegetable or flower gardens. Many tall plants, such as sunflowers and raspberries, can be kept upright by using bamboo canes. Set wooden posts into the soil about a metre (a yard) apart and attach horizontal bamboo canes to these by tying them together with unrottable string. The plants are then tied on to the bamboo canes. This looks much nicer than wire and is also more stable.

It is not only in the garden that bamboo canes are useful. A door curtain made of bamboo and beads not only looks good, it also keeps flies out. The individual internodes should be sawn off at each side of the nodes. If they are left on they have to be drilled through, which is quite hard work. Then bamboo segments and wooden or glass beads are threaded onto strong nylon, in whatever pattern you so desire. These are secured at the top on a drilled piece of wood hung underneath the lintel. The proportion of bamboo and beads can be varied to create beautiful patterns.

If a bamboo curtain is too much like hard work you might consider making a bamboo mobile for a window or door opening. Attach different length bamboo segments to a bamboo cane using nylon thread.

Bamboo can be bent easily after heating, and original and attractive holders for bathroom towels or kitchen cloths can be made in this way. Bend the bamboo over a flame into a large circle and tie the ends together with coloured string, or push the thinner end inside the thicker and stick with wood glue. Hang this on a rope as a holder

A decorative water feature using bamboo culms

Splitting a large bamboo culm

Drilling

If you drill a hole through bamboo, for example to thread a ribbon or to take a screw, always drill from the outside. It is also better to use a metal drill bit rather than a normal wood bit, so as not to split the bamboo. Best of all to use is a drill with a heated bit that burns through. In this way you can drill a hole in a straight line directly through the stem, in any required diameter, and it will not fray.

Sawing

A wood saw is not very suitable for cutting, particularly if the bamboo is thin, and it is better to use a fine hacksaw or, failing that, a very sharp knife.

for hand towels. Another possibility is to stick two or three pieces of cane between two painted pieces of wood as a holder for towels and other cloths.

The above illustrates just a few of the things which can be made using the strong yet supple properties of bamboo and, with a little imagination, there are no limits to its use. Those with a knack for handicrafts will be able to make bird boxes, venetian blinds, and other intricate objects from bamboo canes.

However, always bear in mind that bamboo should be handled differently from wood.

Splitting

For some handicrafts you need split bamboo, not the complete round canes. Short sections can be cut by knocking a knife through the cane using a wooden mallet. For splitting an entire stem it is better to clamp the knife upright in a vice and to

Split bamboo hoops marking the edge of a path

knock the stem against it with soft hammer blows. If you have an old flat iron, split bamboo canes can be ironed flat. Rest them on a smooth fire-resistant surface, heat the iron to very hot and press down hard to iron out the stems.

Bending

Bamboo can be bent when cold or heated. Thin, freshly cut canes can be bent into small circles, and they keep their shape if they are held firmly in position for a few days.

Narrow bamboo splits can be bent easily into circles. These can be sewn together to make chains, for use as curtains or for hanging lamps from. Thicker pieces of bamboo should be bent when warm. They become soft and pliable at about 150° C (330° F) and can be worked into almost any required shape over a gas flame or the heat of glowing charcoal.

Distortion

In Asia bamboos are often distorted to make them rectangular, and this can be done easily with thin culms. The shoots are made to grow through a rectangular pipe (almost like a chimney) so that the soft, growing culms take on a square shape. When it grows out of the top the shape is retained.

Surface preparation

Bamboo culms look very attractive without treating, especially the striped or speckled types, but they look even better when polished with hot wax.

If you are not happy with the natural colour they can be etched or painted. Ferrous sulphate colours the stem black, nitric acid brown and copper sulphate green. Holding bamboo in an open flame brings out brown spots.

Painting

Bamboo culms can be painted, but the surface needs to be primed first, using a fine sandpaper so that the paint will adhere. Household objects are best painted with coloured inks and those for use outside with a waterproof lacquer.

10

Problems with Bamboo Cultivation

Bamboos are not very susceptible to disease or pests. Nevertheless they sometimes do not thrive. In 90 per cent of such cases they have not been looked after correctly and in the remaining 10 per cent of such cases the plant had a problem when it was purchased (see p. 81).

What can actually go wrong with bamboos?

No new growth

If no new growth has appeared two years after transplanting, the rhizome may be dead or seriously damaged. The bamboo is unlikely to recover.

Leaves hanging limp

First make sure that the problem is not simply lack of water, otherwise assume that the harmony between the leaves and the rhizome mass has been upset. The reserves of the rhizome are no longer sufficient to nourish the aerial parts of the plant. In such cases remove two thirds of all the culms, leaving only the youngest. The plant should then recover.

Culms go soft and start to rot

If the culms go soft and rot at the end of the growing period, their nutrition has been disturbed. This often happens with old *Sinarundinaria* plants. The reason is the same as when the leaves hang limp. Remedy: remove all rotten culms and most of the older culms.

Leaves roll up

Plant needs water or is in too sunny a position. The leaves roll up to protect the plant from dessication. The leaves should unroll again as soon as it is watered, or as the sun goes behind a cloud.

Plant does not grow

If the plant does not develop, or even gets smaller, it is not happy with the site. Often such plants are on impermeable soils where the water cannot drain away. Remove the plant, dig out the hole and insert a drainage layer (see p. 83) or transplant to a new site.

New growth is weak

If the new growth is weak, the site is probably unsuitable, or it is possible that the winter was too cold for a newly planted bamboo. Leave the weak stems and wait to see if healthier ones develop next season. If the plant repeats the poor pattern of growth next year it needs transplanting.

Voles eating the shoots

Voles can be dealt with by frightening them with noise, blocking their holes with nasty smelling plants or old fish heads, or by gassing. Mice are very keen on new bamboo shoots and the best way to discourage them is to have a cat.

Problems with bamboos grown in containers

The problem is usually that they are either too dry or too wet. Make sure the soil has not been compacted through the use of hard water. If so, then immediate repotting is necessary. Do not let the roots dry out because then the plant will not be able to take up any more water and the foliage will drop off. Also, make sure the plant does not become undernourished.

Pests of bamboo

If bamboos get too dry in the house or conservatory, and especially if their air is too dry, they can be attacked by red spider-mite, aphids and insects. These can be controlled using standard house plant remedies. Red spider-mite attacks when it is not only too dry but also too warm, and spreads especially quickly in centrally heated rooms. The mites are difficult to spot with the naked eye, but the damage they cause is obvious: the leaves go pale, then yellow and finally dry up. They can be treated with a special acaricide which is applied three times at ten-day intervals, but this is effective against only the larvae and adults,

not against the eggs. If the treatment is not given at these intervals the newly hatched mites will have had chance to lay more eggs and start a new cycle. Above all, treatment has to be varied, otherwise strains develop that are resistant to the acaricide.

Scale insects and aphids affect only those bamboos grown indoors or in conservatories. You often first notice scale insects by the little spots on the underside of the leaves, where each insect has sucked at the leaf, as it sits protected by its hard shield. Aphids cause the leaves to develop a shiny, waxy, sticky covering – so-called honeydew. This layer prevents the leaf from assimilating and if the plant is badly affected it can die. Aphids and scale insects are by no means easy to tackle. Watch closely so that an attack is detected early on. If caught soon enough they can be dealt with by using a solution of soap and spirit. If the plant is badly affected, however, a proprietary chemical spray should be used; making sure that the directions for use are properly followed.

11

Bamboo as a Vegetable

Bamboos are important in the kitchen in all Asiatic countries, and it is the tender young shoots that are eaten. They go with almost all other vegetables and with meat, fish and shellfish, and take on the flavour of the spices used to prepared them. Outside their native areas we usually buy them in tins, but the flavour of these is not comparable with fresh bamboo shoots, just as tinned asparagus is not as good as fresh. In Asia cooked bamboo shoots are sold at market in water and there is a good reason for this. Bamboo shoots dry out very quickly, losing their flavour, and they should therefore be prepared fresh in order to appreciate their excellent flavour. They are also very nutritious, like all seedlings and shoots.

Bamboo shoots cannot be harvested in temperate climates like they are in Asia – as thick as your arm. But you can prepare a tasty dish from garden *Phyllostachys* in the spring, using shoots from unwanted rhizomes. This should be done in the morning, before the soil has warmed up, because the shoots taste at their best if picked at this time.

Shoots of Moso (*Phyllostachys heterocycla* var. *pubescens*) and 'spring rain bamboo' (*P. nidularia*) are harvested early in winter in Japan and China. These early shoots, known as 'winter bamboos' in the kitchen, are a rare delicacy and correspondingly expensive. The soil characteristics affect their flavour, rather in the way that the same grape cultivar produces wine of a different character when grown on a different soil.

All bamboo shoots can, in theory, be eaten, but those of *Arundinaria hindsii*, *Phyllostachys viridis*, *P. nigra* f. *henonis* and *P. heterocycla* taste best. In some areas of France *Sasa kurilensis* is even planted in vegetable gardens. The shoots of this species can be grilled or fried, without previously being boiled. All other bamboo shoots should be boiled, and then the water drained off. Asiatic recipes recommend boiling for an hour or more, but this is necessary only for thick shoots and the asparagus-thin shoots available here need much less time. Freshly gathered shoots can be cooked with their protective sheaths for about 20 to 30 minutes in lightly salted water. Leave them to stand in the water for about 10 minutes after boiling and then remove the sheaths, initially from just one shoot to taste. If it is still bitter cook the shoots for a little longer in fresh water. For some recipes you need to cut the shoots into slices, at an angle; for others they can be left whole.

Recipes

Some gardeners might like to try cooking with bamboos rather than simply appreciating them as they grow. The following recipes are for a few dishes that can be prepared using ingredients readily available in the 'Far-East section' of large food shops.

Hot vegetable salad

Ingredients: 100 g bean sprouts, 100 g bamboo shoots, 50 g small carrots, 100 g celery, 4 dried Tongu fungi, or alternatively mushrooms, 1 green pepper, meat or vegetable stock, 4 dessert-spoons vegetable oil, 1 dessert-spoon each of sesame oil, soya sauce and vinegar, and 1 teaspoon each of salt and glutamate.

The Tongu fungi, available in good delicatessens, should be softened for an hour in warm water. Bean sprouts can be bought fresh or tinned, or you can grow your own.

Cut the carrots, pepper, celery and bamboo shoots into thin slices. Stir-fry the carrots, pepper and bean sprouts in two dessert-spoons of smoking-hot vegetable oil, in an iron pan, for two minutes. Cook the celery and bamboo shoots together in a second pan with the fungi and salt, for about three minutes, stirring frequently. Then mix the contents of both pans together with a little stock, seasame oil, the soya sauce, vinegar and glutamate, and stir vigorously over a high heat for two minutes. This dish can also be eaten cold.

Cold vegetable salad

Ingredients: 100 g bamboo shoots, 100 g yellow savoy cabbage leaves, 100 g lettuce, 3 young carrots, 3 tomatoes, 3 spring onions or shallots, 1 large onion, a few radishes, 2 cloves garlic, 2 slices root ginger, 2 dessert-spoons each of soya sauce and vinegar, half a teaspoon of chilli sauce, 3 teaspoons sesame oil, a little glutamate, chopped chives and chopped green coriander.

Cut the savoy cabbage, carrots, radishes and lettuce into thin slices, boil the bamboo shoots and rinse in cold water and cut the tomatoes into eighths. Chop the spring onions into 2 cm (1 in.) lengths, or, if shallots are used, peel and halve them, slice the onion into rings and crush the garlic with salt. Heat the oil in an iron pan and cook the onions, ginger and garlic over a medium heat for three minutes, stirring frequently. Then pour onto a plate to cool the oil quickly.

Put all the other ingredients into a large bowl, add the by now lukewarm onions and the sesame oil, salt, glutamate, soya sauce, vinegar and chilli sauce. Mix well and finally garnish with the chives and coriander.

Double-cooked bamboo shoots

Ingredients: 500 g fresh bamboo shoots, 75 g Chinese pickles (these can be bought as 'red in snow' in Asian countries), 5 dessert-spoons vegetable oil, 2 dessert-spoons each of soya sauce and vegetable stock (cube), 1 teaspoon each brown sugar and sherry, a little glutamate.

The bamboo shoots are used raw in this recipe.

Wash the bamboo shoots, peel off the sheaths and cut the soft inner part into oblique thin slices. Chop the Chinese pickles and wash in hot water. Heat 2 dessert-spoons of vegetable oil in a pan and stir in the pickles until they are crisp, then drain in a sieve. Fry the bamboo shoots in the remaining oil over medium heat until they are coated in oil, taking care not to damage the delicate slices. Drain after about five minutes when they are golden-brown.

Now put the pickles back in the pan, add the bamboo shoots, and mix in the sugar, sherry, soya sauce, glutamate and vegetable stock.

Stir over a high heat for a minute, then serve immediately.

Basted bamboo shoots

250 g bamboo shoots, 25 g chives, frying oil, salt, 2 teaspoons sesame oil.

In Asia people tend to take a long time over their cooking and time-consuming recipes are common. The piquant flavour and attractive presentation of the food are equally important. This recipe is an example.

First boil the shoots and lay them lengthways as a thin layer in a wire basket. Cut the chives to the same length and lay them in the same direction. Slowly spoon boiling oil over these to baste them, repeating the basting 20 times. Then transfer to a dish, making sure that the contrasting colours of the bamboo and chives are well displayed, sprinkle with salt and pour on the sesame oil. Serve very hot, with meat or fish.

Bamboo shoots, Szechuan style

Ingredients: dry Tongu fungi, 500 g bamboo shoots, 75 g bean sprouts, 100 g pork belly, 2 spring onions or shallots, 4 dessert-spoons pork fat, 3 dessert-spoons soya sauce, 1 teaspoon each of chilli sauce and sugar, 1 dessert-spoon cornflour, 250 ml chicken stock.

Boil the bamboo shoots, peel them and cut obliquely into thin slices. Cut the pork into thin strips and the spring onions and bean sprouts into small pieces.

Heat the pork fat in a large pan, add the bamboo shoots and gently stir for four minutes. Remove the shoots and keep them warm. Then fry the pork in the pan for one minute at a high heat. Add the onions, bean sprouts and softened fungi and stir vigorously. Then return the bamboo shoots to the pan, adding the soya sauce, chilli sauce and sugar. Cook over a gentle heat for three minutes. Meanwhile mix the cornflour smoothly into the cold chicken stock and add to the pan. The dish is ready after a further three minutes gentle cooking.

Bamboo shoots with fungi: Vietnamese-style

Ingredients: 250 g bamboo shoots, 12 dry Tongu fungi or others, 2 pieces of onion stem, 1 small piece of root ginger, 1 large onion, 1 bunch of peppermint, 1 bunch of green coriander, 2 dessert-spoons oil, 1 clove of garlic, 1 teaspoon chilli sauce, 4 dessert-spoons oil, 1 clove of garlic, 1 teaspoon chilli sauce, 4 dessert-spoons stock and 6 dessert-spoons *Nuc-mam* sauce.

Boil the bamboo shoots, peel and cut into thin strips. Soften fungi for at least 30 minutes in warm water and cut them into strips. Slice the ginger and onions into thin strips.

Lightly fry the onion, crushed garlic and salt in holt oil. Then add the fungi and stir together for three minutes. Then add the stock, pepper, chilli sauce, ginger and the *Nuc-mam* sauce and cook, stirring, for one minute.

Put the shoots on a plate and garnish with a chopped onion stem, green coriander and peppermint (the leaves should be eaten). Serve the sauce separately. This dish should be served with boiled rice.

Bamboo shoots with fungi: Chinese-style

Ingredients: 200 g bamboo shoots, 200 g fungi, 1 dessert-spoon Chinese morels, 1 teaspoon cornflour, 2 dessert-spoons each of vegetable oil and sesame oil.
For the sauce: 5 dessert-spoons chicken stock,

1 dessert-spoon soya sauce and rice wine, 1 pinch each of sugar, ginger and salt.

Boil, peel and slice the bamboo shoots and soften the morels. Fry the fungi and bamboo shoots together for one minute, at a high heat, stirring constantly. Mix the sauce ingredients together and add to the pan, cooking for a further two minutes. Thicken with a little cornflour, remove from the heat and sprinkle with sesame oil.

Bamboo shoots with roe

Ingredients: 400 g bamboo shoots, 100 g fish roe (tinned), 1 dessert-spoon soya sauce, 3 dessert-spoons rice wine, 1 teaspoon cornflour, a pinch each of salt and sugar.

Boil, peel and slice the bamboo shoots, rinse the roe (assuming it is salted) and stand it in ginger powder for about 10 minutes. Fry the bamboo shoots for two minutes in a hot pan, and the roe for about half a minute. Add the soya sauce and rice wine, thicken with cornflour, season and serve.

Bamboo shoots with pork

Ingredients: 200 g bamboo shoots, 200 g lean pork, 200 g fresh asparagus, 1 dessert-spoon finely chopped shallots, 1 teaspoon finely chopped ginger, 1 teaspoon sesame oil, 1 teaspoon cornflour, salt, pepper, oil for frying.
For the sauce: 1 dessert-spoon each of soya sauce and rice wine or sherry, 1 teaspoon each of vinegar, sugar and cornflour.

Slice the pork thinly, mix with soya sauce, cornflour and sesame oil and stand for 20 minutes. Peel the asparagus and cook for 20 minutes, boil the bamboo shoots and remove the leaves. Cut both into bite-sized pieces. Mix the asparagus water with the sauce ingredients. Fry the shallots in oil in an iron pan, then add the pork and fry for two minutes, stirring throughout. Only then add the bamboo shoots and fry for a further one minute. Lower the heat and add the sauce, cooking for a short time, then right at the end add the asparagus and season with salt and pepper. The same dish can be prepared using slices of chicken breast.

Chicken with bamboo shoots

Ingredients: 300 g chicken meat, 150 g bamboo shoots, 1 red and 1 green chilli pod, 1 dessert-spoon each of peppercorns, rice wine and plant oil, sugar, salt, oil for frying.
For the marinade: 1 egg white, 1 teaspoon each of cornflour and rice wine, salt.

Cut the chicken into thin strips and put into a marinade of 1 egg white, 1 teaspoon cornflour, 1 teaspoon rice wine, and a little salt. Leave this in the fridge for half an hour. Meanwhile boil, peel and slice the bamboo shoots, and the chilli, removing the seeds if you do not want the dish too hot.

Fry the marinaded chicken briefly, just until it turns white. Fry the bamboo shoots in oil in a frying pan with a pinch of salt for one minute. Remove and place them on a plate. Add the peppercorns, one dessert-spoon plant oil, the chicken and the chilli to the same pan, fry for two minutes and then add the bamboo shoots and rice wine. Season with salt and sugar and serve very hot.

Appendix

Bamboos in botanic gardens and parks

Bamboos are some of the most photogenic of all plants. The beauty of a bamboo comes not simply from the elegant shape and colour of the culm and leaves, and the overall harmony of the whole plant. Rather, their true fascination derives from the interplay of light and shade produced by their constantly moving foliage. You do not need to travel to Asia or South America to see bamboos – they can also be found increasingly in botanic gardens. Whereas only a decade or so ago there were only a few examples to be seen, today there are dozens of different genera and species represented. Whoever has the opportunity, and an interest in bamboos, should certainly take a 'bamboo walk' in a local botanic garden. The sight of a well-grown bamboo can also be an inspiration to those planning to integrate bamboos into their own garden. The reader will find here a list of botanic gardens with good displays of bamboos. This list cannot however be comprehensive because more and more botanic gardens are now planting bamboos, and there are even new specialist bamboo gardens as well. An attempt to list all the genera and species found in these would founder on taxonomic difficulties. Moreover, the older established plantings are often not well labelled and require fresh identification. The reader therefore needs to be aware of these taxonomic problems when visiting bamboo collections.

Canada
Buchart Garden
Victoria
BC

France
Bambuserai de Prafrance
30140 Anduze

Germany
University Botanic Garden
Bonn

Botanic Garden and Botanical Museum
Berlin-Dahlem

University Botanic Garden
Erlangen

Gruga-Park
Essen

Palmengarten
Frankfurt

New Botanic Garden
Hamburg-Flottbeck

University Botanic Garden
Karlsruhe

Bodensee
Mainau Island

University Botanic Garden
Oldenburg

Botanic Garden
Nympenburg
Munich

Great Britain
Royal Botanic Gardens
Kew
Richmond
Surrey
TW9 3AB

USA
USDA Plant Introduction Station
Coral Gables
FLA

Fairchild Tropical Garden
10901 Old Cutler Road
Miami
FLA 33158

Kanapaha Botanical Gardens
4625 SW 63rd Blvd
Gainesville
FLA 32601

Coastal Extension Research Center
Georgia Agriculture Extension
P.O. Box 9866
Savannah
GA

Lyon Arboretum
3860 Manoa Road
Honolulu
HI 96822

Jungle Gardens
Avery Island
LA

Longwood Gardens
Kennett Square
PA 19348

Suppliers

Nurseries have recently started stocking more and more bamboos. It is not possible here to list all the firms involved, only certain ones which have specialized, such as the following:

France
Bambouseraie de Prafrance
F-3104 Anduze/Gard

Germany
Baumschule Wolfgang F. Eberts
Saarstrasse 3
D-7570 Baden Baden

Sortiments- und Versuchsgärtnerei Hans und Helga Simon
Staudenweg
D-8772 Marktheidenfeld

Baumschulen Gertrud Willumeit
Nussbaumallee 69
D-6100 Darmstadt

Italy
Firma Baldacchi
Pistoia

Switzerland
Firma Vatter
CH-3098
Köniz bei Bern

United Kingdom
Bamboo Nursery
Kingsgate Cottage
Wittersham
Tenterden
Kent
TN30 7NS

Drysdale Nursery
Bowerwood Road
Fordingbridge
Hampshire
SP6 1BN

Jungle Giants
Plough Farm
Wigmore
Herefordshire
HR6 9UW

USA
Bamboo Sourcery
666 Wagnon Road
Sebastopol
CA 95472

Bamboo Societies

The increasing number of bamboo enthusiasts has led recently to the formation in France of the European Bamboo Society (EBS). Member countries are Germany, France, United Kingdom, Italy and Switzerland.

The aim of the EBS is to spread knowledge about bamboos. It has also fostered the cultivation of rare and endangered bamboo species and varieties. For further information contact the society at the following addresses:

Australia
c/o Mr Hans Erken
P.O. Box 500
Maleny
Queensland 4552

Canada
Bamboo Newsletter of Canada
c/o Michael Curtis
14050 – 60th Avenue
Surrey B.C.
V3X 2N3

France
EBS
c/o Yves Crouzet
Bambouseraie de Prafrance
Générargues
30140 Anduze

Germany
EBS
c/o Roland Eitel
Feldstrasse 37
6466 Gründau 2

Great Britain
EBS
c/o David Helliwell
43 Whitehouse Road
Oxford
OX1 4QJ

Italy
EBS
c/o Bar Lorenzo
Borgata Mascarelli 47
12064 La Morra

The Netherlands
EBS
c/o Charley Younge
Dorpsweg 125
1697 KJ Schellinkhout

New Zealand
The New Zealand Bamboo Society
c/o Mrs Nicki Higgie
P.O. Box 11
Fordell
Wanganui

Spain
Asociacion Espãnola del Bambu
c/o José Maria Viure
Bambuseria Viure – Bonsai Viure C.B.
Carretera Cardedeu a Cánoves, km. 2 izq.
08440 Cardedeu

Switzerland
EBS
c/o Toni Grieb
Montet
1588 Cudrefin

USA
American Bamboo Society
c/o Gerald Bol
666 Wagnon Road
Sebastopol
CA 95472

Bibliography

AUSTIN, R. und K. UEDA: Bamboo. Verlag Weatherhill, New York o.J.

BEUCHERT, Marianne: Die Gärten Chinas. Verlag Eugen Diederichs, Frankfurt 1983.

BUT, Paul Pui-Hay, Liang-chi CHIA, Hok-lam FUNG und Shiu-Ying HU: Hong Kong Bamboos. The Urban Council Hong Kong 1985.

CROUZET, Yves: Les Bambous. Selbstverlag Yves Crouzet, Prafrance 1984.

DUNKELBERG, Klaus: Bambus als Baustoff. Koldewey-Gesellschaft 1980.

FARRELLY, D: The Book of Bamboo. Sierra Club Books, San Francisco 1984.

KONISHO, Kiyoko: Japanisch kochen hält fit und gesund. Kikkoman, Düsseldorf 1985.

KUHKAUPT, Dieter: Gärten Japans. Verlag Du Mont, Köln 1985.

LAWSON, A.H.: Bamboos, A Gardener's Guide to their Cultivation in Temperate Climates. Taplinger Publishing, New York 1968.

LO, Kenneth: Das große Buch der chinesischen Kochkunst. Econ-Verlag, 1980

McCLURE, F.A. The Bamboos. A fresh perspective. Harvard Univ. Cambridge Mass. 1966.

SUZUKI, S.: Index to Japanese Bambusaceae Gakken Co., Ltd. Tokyo, Japan.

TAKAMA, Shinji: Die wunderbare Welt des Bambus. Verlag Du Mont, Köln, 1983.

ULENBROK, Jan: Haiku, Japanische Gedichte. Wilhelm Heyne Verlag, München 1985.

YOUNG, Robert: Bamboo in the United States. USDA-publication 1961.

Journals

Bambusblätter, Herausgeber Werner Simon, D-8772 Marktheidenfeld, Staudenweg.

Newsletters and Yearbooks of the American Bamboo Society, P.O. Box 640, Springville, CA 93265, USA.

The Bamboos of the World, Fortsetzungswerk. Verlag D. Ohrnberger, Wisenstr. 5, D-8901 Langweid a. L., Germany.

Picture acknowledgements

All photographs by Max Felix Wetterwald, Offenburg, except the following:

Helmuth Flubacher, Fellbach, prepared the drawings.

The rhizome illustrations are after D. Farelly and F.A. McClure.

The distribution map on page 23 reproduced by courtesy of Josef Goerings and Dieter Ohrnberger.

General Index

Botanical Index

127